Principles of Renal Physiology

4th Edition

Principles of Renal Physiology

4th Edition

Chris Lote

Professor of Experimental Nephrology,
University of Birmingham, UK

Kluwer Academic Publishers
Dordrecht Boston London

A C.I.P. Catalogue record for this book is available from the Library of Congress

ISBN (hardback) 0-7923-6074-5
ISBN (paperback) 0-7923-6178-4

Published by Kluwer Academic Publishers,
P.O. Box 17, 3300 AA Dordrecht, The Netherlands

Sold and distributed in North, Central and South America
by Kluwer Academic Publishers,
101 Philip Drive, Norwell, MA 02061, U.S.A.

In all other countries, sold and distributed
by Kluwer Academic Publishers Distribution Centre,
P.O. Box 322, 3300 AH Dordrecht, The Netherlands

Printed on acid-free paper

Printed and bound in Great Britain by Anthony Rowe Limited

Contents

Preface to the fourth edition

The first edition of this book appeared in 1982. In the preface to that first edition, I wrote 'This book is based on the lecture course in renal physiology which I give to medical students at the University of Birmingham. The purpose of the book is primarily to set out the principles of renal physiology for preclinical medical students, and it is therefore concerned mainly with normal renal function. However, diseases or abnormalities in other body systems may lead to adaptations or modifications of renal function, so that a good knowledge of renal physiology is essential to the understanding of many disease states, for example the oedema of heart failure or liver disease, or the consequences of haemorrhage and shock.' The new edition is still based on the lectures which I continue to give at Birmingham University, but over the years the course has gradually changed, to being a system based course covering all aspects of the kidney – the anatomy, physiology, pharmacology and pathology. The new edition of the book, which has been extensively revised and rewritten, reflects this. However, it continues to offer a concise, easily readable format, primarily intended for undergraduate medical and medical science students. This new edition will carry the book through to twenty years of continuous publication, so it must be doing something right! I hope the new generation of readers will find it interesting and that it meets their needs, and I would like to encourage all of them to let me have their comments and suggestions.

Chris Lote

Terminology and abbreviations

General

ADH	Antidiuretic hormone (vasopressin)
ANP	Atrial natriuretic peptide
ECG	Electrocardiogram
ERPF	Effective renal plasma flow
GFR	Glomerular filtration rate
RBF	Renal Blood flow
RPF	Renal plasma flow

Units of measurement

mosmol	milliosmoles
mmHg	millimetres of mercury (pressure measurement)
mM	millimoles/litre
mmol	millimoles
nm	nanometres (10^{-9} metres)
μm	micrometres (10^{-6} metres)

Symbols

Square brackets, [], denote concentration, e.g. plasma $[Na^+] = 140\,mM$ means plasma sodium concentration $= 140\,mM$.

The body fluids · 1

1.1 *Introduction*

The body fluids can be considered to be distributed between two compartments, intracellular and extracellular. The extracellular compartment can in turn be divided into a number of sub-compartments. These are: (a) the plasma (extracellular fluid within the vascular system); (b) the interstitial fluid (extracellular fluid outside the vascular system, and functionally separated from it by the capillary endothelium); and (c) transcellular fluids. Transcellular fluid can be defined as extracellular fluid which is separated from the plasma by an additional epithelial layer, as well as by the capillary endothelium. Transcellular fluids have specialized functions and include the fluid within the digestive and urinary tracts, the synovial fluid in the joints, the aqueous and vitreous humours in the eye and the cerebrospinal fluid.

The environment in which our cells exist is not the environment of the outside of the body. The immediate environment of the cells is the extracellular fluid. This **internal environment** (a term first used by the nineteenth century French physiologist, Claude Bernard) provides the stable medium necessary for the normal functioning of the cells of the body. The internal environment maintains the correct concentrations of oxygen, carbon dioxide, ions and nutritional materials, for the normal functioning of the cells.

Maintenance of the constancy of the internal environment (which the American physiologist, Walter B. Cannon, called 'homeostasis') is ensured by several body systems. For example, the partial pressures of oxygen and carbon dioxide are regulated by the lungs. The kidneys play a vital role in homeostasis. In fact, they can reasonably be regarded as the most important regulatory organs for controlling the internal environment, since they control not only the concentration of waste products of metabolism, but also the osmolality, volume, acid–base status and ionic composition of the extracellular fluid and, indirectly, regulate these same variables within the cells.

Much of this book is devoted to an examination of the ways in which the kidneys perform these regulatory functions. First, however, it is necessary to examine the normal composition of the body fluids.

1.2 *Body water*

Water is the major component of the human body and, in any individual, body water content stays remarkably constant from day to day. However, there is considerable variability in the water content of different individuals and this variability is due to differences in the amount of adipose tissue (fat) in different people.

In a 70 kg man of average build, the body water will constitute 63% of the body weight, and thus there will be 45 l of total body water (TBW). In a woman of the same weight, only about 52% (36 l) of the body weight will be water. This difference is due to the fact that women have more adipose tissue than men and the water content of adipose tissue is very low (about 10%).

In obese people, fat is a major constituent of the body (second only to water), and even 'slim' people have considerable quantities of fat. We can regard the fat as non-functional (storage) tissue. The functional tissue of the body can be regarded as fat-free: the percentage of water in the fat-free tissue is extremely constant, both within an individual from day to day and between individuals. The percentage water in this 'lean body mass' is 73%.

1.3 *Body fluid osmolality*

The exchange of water between the different body fluid compartments is facilitated by two forces: hydrostatic pressure and osmotic pressure. In order to understand osmotic pressure, consider a container of water, separated into two compartments by a membrane permeable to water. The water molecules will be moving at random (Brownian motion) and some of them will be moving across the membrane by diffusion; the rate of diffusion in each direction will be equal, so that net flux is zero. Now, suppose that a solute is added to one compartment. The addition of solutes to water reduces the random movement (activity) of the water molecules and, consequently, the diffusion of water from the side containing the solute to the side containing only water will be reduced. There will then be a net flux of water from the pure water side to the solution side of the membrane (Figure 1.1). We can measure this osmotic effect as an **osmotic pressure**, by determining the hydrostatic pressure which must be applied to the compartment containing solute, to prevent the net entry of water. This hydrostatic pressure is equal to the osmotic pressure of the solution.

1.3.1 *Units of osmotic measurement*

From the foregoing, it is apparent that, like hydrostatic pressure, osmotic pressure could be expressed as mmHg. A more useful unit, however, is the osmole, and in physiology osmolality is usually expressed as mosmol/kg H_2O. The

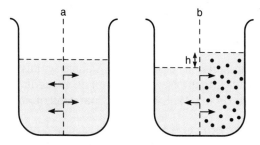

Figure 1.1 A container divided into two compartments by a membrane permeable to water, but impermeable to some solutes. (Such membranes are termed 'semi-perme-able'.) (a) When there is only water in the container, the unidirectional water fluxes (represented by the arrows) are equal, so that the net flux of water is zero. (b) If a solute to which the membrane is impermeable is added to one compartment of the container, the activity of the water molecules on the side containing the solute is reduced, so that the unidirectional water flux from that side is reduced. The unidirectional water flux into the solute-containing side continues as before, so that there is a net flux into the solute-containing side, which creates a hydrostatic pressure difference (*h*). This hydrostatic pressure is equal to the **osmotic pressure** of the solution. Alternatively, the osmotic pressure of the solution could be measured by determining the hydrostatic pressure which must be applied to the solute-containing side to prevent the net influx of water to this compartment. Water moves from an area of *low* osmolality to one of *higher* osmolality.

osmole is analogous to the mole (and, for non-dissociable substances, is *identical* to the mole, i.e. it is 1 g molecular weight of the non-dissociable molecule). For molecules which dissociate, each particle derived from the molecule contributes to the osmotic pressure and so, to calculate the osmolality from the molality, it is necessary to multiply by the number of particles. Suppose a molecule dissociates into *n* ions; then osmolality is given by

$$\text{Osmolality (mosmol/kg } H_2O) = n \times \text{molar concentration (mmol/kg } H_2O)$$

It should be noted that the units of osmolality (and molality) refer to the concentration per unit weight of *solvent* (water). This is in contrast to osmolarity (and molarity) which refers to the (os)molar concentration per litre of solution (i.e. water + solute), so units of osmolarity are mosmol/l. Osmolality is the preferred measurement, although in practice the difference between the two terms when dealing with physiological solute concentrations is very small.

1.3.2 *Isotonicity and isosmoticity*

If red blood cells are suspended in distilled water, water enters the cells by osmosis and the cells swell and burst. This is haemolysis. If the cells are

Osmolality and osmolarity

Osmotic concentrations in biology are usually measured in units of osmolality – typically mosmol/kg H_2O. What do these units mean? Suppose we make up a 1 MOLAR solution of glucose…

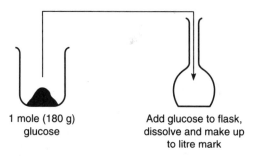

| 1 mole (180 g) glucose | Add glucose to flask, dissolve and make up to litre mark |

We take 1 mol of glucose (180 g), put it in a 1 litre flask, add some water to dissolve it, then add more water to make it up to the litre mark.

What we have is a 1 molar solution. The glucose concentration is 1 mol/l, i.e. 1000 mmol/l. (These are units of MOLARITY.)

Since glucose does not dissociate, the number of osmotically active particles is equal to the number of glucose molecules. So the osmotic concentration (=total particle concentration) is 1 osmol/l, i.e. 1000 mosmol/l. (These are units of OSMOLARITY.)

But we do not know how much water there is in the flask (because we do not know the volume occupied by the glucose). So the units are **per litre of solution**.

Suppose that, instead of glucose, we weigh out 1 mol (58.5 g) salt (sodium chloride, NaCl), and make it up to 1 litre as above. The NaCl concentration is 1 mol/l = 1000 mmol/l.

But NaCl dissociates into Na^+ ions and Cl^- ions, so the number of osmotically active particles is almost double the number of NaCl molecules (**almost, not exactly**, double, because the dissociation is not complete). The total particle concentration of a 1 mol/l NaCl solution is 1.86 osmol/l. (1860 mosmol/l.) As above, these are units of OSMOLARITY.

Again, we do not know how much water there is in the flask, because we do not know the volume occupied by the solute. But there is another way of making standard solutions…

(*Continued*)

If instead of putting the solute (e.g. glucose) into the flask, and adding water, we fill the flask up to the litre mark with water (i.e. 1 kg water) and then add this to the 1 mol glucose in the beaker, the solution is 1 mol glucose per kg H_2O.

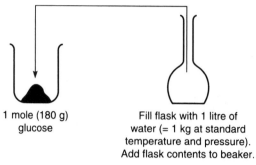

1 mole (180 g)
glucose

Fill flask with 1 litre of
water (= 1 kg at standard
temperature and pressure).
Add flask contents to beaker.

This is called a 'MOLAL' solution. Since the number of osmotically active particles in it is the same as the number of glucose molecules, a solution which is 1 mol glucose/kg is also 1 osmol/kg H_2O, i.e. 1000 mosmol/kg H_2O. These are units of OSMOLALITY.

If we were to make up a sodium chloride solution in the same way, using 1 mol NaCl, the solution would have an osmolality of almost 2000 mosmol/kg H_2O (actually 1860 mosmol/kg H_2O). These are units of OSMOLALITY. The body fluid osmolality is 285 mosmol/kg H_2O.

Reproduced with permission from C.J. Lote *Biol. Sci. Rev.* (1993) **5** (3), 20–23.

suspended in a solution which does not cause any change in the cell volume, such a solution is termed **isotonic**. The most widely used isotonic solution is 0.9% saline; 5% dextrose is also isotonic. Both of these solutions have an osmolality of about 285 mosmol/kg H_2O, as do the cell contents and the plasma. Solutions having the same osmolality are termed **isosmotic**. However, although different isotonic solutions are isosmotic, solutions which are isosmotic to the plasma are not necessarily isotonic. This can be demonstrated if red cells are suspended in an isosmotic urea solution. The cells swell and haemolyse, just as they do in distilled water. This occurs because urea can readily cross cell membranes and thus equilibrates across the membrane to reach the same concentration on each side, so that the osmotic effect of the urea is cancelled and the osmolality of the other intracellular solutes causes the cell to swell (Figure 1.2).

Whether a solute is an ineffective osmole in this way depends not only on the solute, but also on the properties of the membrane. Thus urea is an ineffective osmole across the cell membrane, because it diffuses so rapidly, whereas

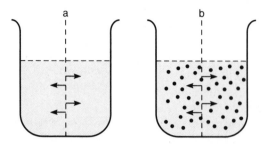

Figure 1.2 A container divided into two compartments by a semi-permeable membrane, as in Figure 1.1. (a) Net flux of water is zero, as in Figure 1.1a. (b) Addition to the compartment on the right of a solute which can readily cross the membrane does not significantly alter the net water flux, because the solute rapidly equilibrates across the membrane (i.e. reaches the same concentration on each side). A solute which behaves in this way is termed an 'ineffective osmole'.

Na^+ is an effective osmole. In contrast, the capillary membrane is much more permeable than the cell membrane, so *all* the solutes in the plasma except the plasma proteins are ineffective osmoles in relation to the capillary walls. An effective osmole is a solute which, when placed on one side of a semi-permeable membrane, will tend to cause water movement across that membrane. An ineffective osmole is one which, when placed on one side of a semi-permeable membrane, will itself rapidly diffuse across the membrane.

1.3.3   *Normal body fluid osmolality*

The body fluid osmolality is maintained remarkably constant, at 280–290 mosmol/kg H_2O. Chapter 8 describes how this constancy is achieved. It should be noted that, in general, intracellular and extracellular fluid osmolalities are identical, because water can readily cross the cell membranes.

1.4   *Distribution of ions across biological membranes*

Because of their random thermal motion (Brownian movement), the individual molecules of dissolved substances are continually diffusing, so that there are no concentration differences between different parts of a solution. (If the concentration of a dissolved substance is higher in one region than another, then molecules will diffuse in both directions, but more will diffuse from the region of high concentration to the region of lower concentration than vice versa, so that the concentrations become equal.)

Biological membranes (e.g. cell membranes, capillary membranes) permit the diffusion of small molecules and ions through them. However, substances diffuse through biological membranes (particularly cell membranes) much more slowly than they diffuse freely in water, and different molecules and ions diffuse through membranes at different rates depending on their molecular weight, shape and charge. So the cell membrane is *selective*, allowing some molecules to pass much more readily than others. Some large molecules (e.g. proteins) diffuse across the cell membrane so slowly that the membrane is effectively impermeable to them.

The presence of such non-permeant ions (e.g. proteins, which are generally anions at physiological pH) on one side of a membrane which is permeable to other ions, causes the unequal distribution of the diffusible ions – this is the **Gibbs–Donnan effect**. Ion distribution will be such that the products of the concentrations of diffusible ions on the two sides will be equal, i.e. if we call the two sides a and b:

$$\begin{array}{ccccccc}
\text{Diffusible} & & \text{Diffusible} & & \text{Diffusible} & & \text{Diffusible} \\
\text{cations} & \times & \text{anions} & = & \text{cations} & \times & \text{anions} \\
\text{(side a)} & & \text{(side a)} & & \text{(side b)} & & \text{(side b)}
\end{array}$$

A consequence of this distribution is that the total number of ions on the side containing non-diffusible ions will be slightly greater than the total number of ions on the side containing only diffusible ions; so that the osmotic pressure will be slightly greater on the side containing the non-diffusible ions.

Thus there is a slight imbalance of ions between the inside and outside of capillaries, due to the presence of non-diffusible plasma proteins within the capillaries (p. 10).

Across cell membranes, however, the situation is more complicated. Cell membranes are permeable to K^+ and Cl^-; but much less permeable to Na^+ (Na^+ permeance is less than $1/50$ that of K^+). In addition, there are active transport processes in the cell membrane, which actively extrude sodium and keep the intracellular sodium concentration very low. The active Na^+ pump also pumps K^+ into the cell, and since the pump uses ATP, it is a Na^+, K^+-ATPase. However, the permeability of the membrane to K^+ is so high that the pump has little direct effect on the K^+ distribution, whereas it has a very marked effect on Na^+ distribution.

Inside cells, there are proteins which have a net negative charge (protein anions). These will tend to cause the accumulation of positive ions inside the cell. Since Na^+ is actively kept out of the cell, K^+ enters. This makes the K^+ concentration inside very high and there will be some diffusion out down the concentration gradient, so that electrical neutrality will not be achieved and at equilibrium the cell will have a net negative charge (about $-70\,mV$). Chloride will distribute passively across the cell membrane and so, because of the net negative charge inside the cell, there will be more chloride outside. Sodium is present in high concentration outside the cell because of the active sodium pump.

We might expect, as a result of the Gibbs–Donnan effect detailed above, that there would be more ions inside the cell than outside, because of the non-diffusible intracellular proteins. However, these are balanced by sodium outside, which, because of the sodium pump, is effectively non-diffusible. If the sodium pump is chemically inhibited, sodium is no longer effectively non-diffusible and so cells swell by osmosis. Thus the Na^+ pump is essential for the osmotic stability of cells.

Passive movement of solute can occur by a mechanism other than diffusion. This alternative is solvent drag, which is movement (bulk flow) of solvent (water) carrying with it the solutes dissolved in it. Bulk flow occurs in vessels (blood vessels, renal tubules, lymph vessels), but also takes place across membranes, as filtration (see below).

1.5   *Fluid exchanges between body compartments*

Although the body fluid compartments have a relatively constant composition and are in equilibrium with each other, this equilibrium is not static, but dynamic. There is a continual internal fluid exchange, between the plasma and the interstitial fluid, and between the interstitial fluid and the cells.

1.5.1   *Exchanges between interstitial fluid and intracellular fluid*

Cell membranes have a much lower permeability than capillary endothelia to ions and water-soluble molecules. Nevertheless, water can freely cross cell membranes, so that intracellular and extracellular fluids are in osmotic equilibrium. Changes in the ionic content of the intracellular or extracellular fluid will lead to corresponding movements of water between the two compartments. Na^+ ions are the most important extracellular osmotically active ions, whereas K^+ ions are most important intracellularly. If we added more sodium to the extracellular fluid, acutely, e.g. by ingesting NaCl, the extra solute would osmotically attract water from the intracellular fluid until the osmolality of intra- and extracellular fluids was again equal, and the osmolalities of both compartments would be increased, even though the added solute was confined to the extracellular fluid.

1.5.2   *Exchanges between the plasma and interstitial fluid*

The functionally important part of the vascular system, for this exchange to occur, is the capillaries (because the capillaries are the only part of the vascular system with permeable walls). Water and electrolytes move continuously through the capillary walls between the plasma and the interstitial fluid, in both directions. Most of this movement (90%) occurs as a result of simple diffusion. In addition, there is a 10% contribution to the movement by filtration: at the arteriolar end of a capillary, there is a gradient (of hydrostatic pressure) causing fluid filtration through the capillary wall from the vascular system into the interstitial fluid. At the venous end of the capillary, the filtration gradient (due to osmotic pressure) is directed *into* the capillary. This is illustrated in Figure 1.3. (In fact real capillaries, unlike the idealized capillary shown in the diagram, are perfused intermittently: at times when blood is flowing through the capillary, the hydrostatic pressure inside is high enough to cause transudation out of the capillary along its entire length. When flow stops, reabsorption occurs along the whole length of capillary. The overall effect is as if the capillaries behaved like the idealized capillary in Figure 1.3.)

The movement of fluid across the capillary endothelium in each direction has been estimated to be about 120 l/min. Since the plasma volume is only 3 l, it follows that the plasma water completely exchanges with the interstitial fluid every 1.5 seconds!

The diffusive flow is mainly dependent on the physical characteristics of the capillary endothelium, and the size and chemical nature of the solutes. Oxygen and carbon dioxide are lipid-soluble and can therefore diffuse freely across the entire surface of the capillaries (since the cell membranes of the capillary endothelial cells, like all cell membranes, are composed mainly of lipid). Water, small molecules and ions pass through spaces (pores) between the capillary endothelial cells. Of the plasma constituents, only the plasma proteins are unable to cross the capillary walls, so that the plasma proteins exert the osmotic effect across the capillary walls (i.e. are effective osmoles), which is responsible for the re-entry of fluid at the venous end of the capillaries to balance the fluid filtered out by the hydrostatic pressure gradient at the arterial end.

Let us examine the forces governing movements across the capillary wall in more detail.

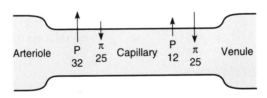

Figure 1.3 Fluid exchange between the plasma and interstitial fluid across the capillary wall. At the arteriolar end (on the left) the capillary hydrostatic pressure (*P*, 32 mmHg) exceeds the osmotic pressure due to the plasma proteins (π, 25 mmHg), so that there is a loss of fluid from the capillary. At the venous end (on the right) the plasma protein osmotic pressure (π, 25 mmHg) is greater than the hydrostatic pressure (*P*, 12 mmHg), because hydrostatic pressure decreases along the capillary. Thus at the venous end, fluid re-enters the capillary.

The **hydrostatic pressure** within the capillary is about 32 mmHg at the arterial end, and has fallen to about 12 mmHg by the venous end. This pressure depends on:

(1)   the arterial blood pressure;
(2)   the extent to which the arterial pressure is transmitted to the capillaries – i.e. the arteriolar resistance;
(3)   the venous pressure.

Normally, the pressure at the arterial end of the capillaries is closely regulated and stays virtually constant. However, alterations in venous pressure will produce changes in capillary hydrostatic pressure.

The **oncotic pressure** (plasma protein osmotic pressure) in the capillary is commonly stated to be 25 mmHg. In fact, the plasma proteins are responsible for there being an osmotic pressure difference of 25 mmHg between the capillaries and the interstitial fluid, but it is not strictly true to say that the pressure is entirely due to plasma proteins. The plasma proteins exert an osmotic effect of about 17 mmHg. However, they have a net negative charge and, since the capillary endothelium is a semipermeable membrane, the plasma proteins cause an imbalance of diffusible ions across the capillary wall, so that there are more ions (mainly sodium) inside the capillary than outside. This imbalance is responsible for a further 8 mmHg osmotic pressure, making a total osmotic pressure of 25 mmHg.

In the interstitial fluid, there are usually small amounts of plasma proteins which have escaped from the capillary. These make little contribution to the balance of forces across the capillary endothelium, but this situation may change if the permeability of the capillaries to proteins increases for any reason, or if the removal of proteins from the interstitial fluid by the lymphatic system is reduced (see below).

It is very difficult to measure the interstitial fluid hydrostatic pressure (sometimes called the tissue turgor pressure). Estimates vary from a figure of 4 mmHg, to about −8 mmHg; it seems reasonable to regard the tissue hydrostatic pressure as approximately zero.

1.5.3   *Lymphatic system*

The lymphatic system is a network of thin vessels (resembling veins) which begins as lymphatic capillaries in almost all the organs and tissues of the body and which eventually drains into the venous system in the neck. Although the lymph capillaries are blind-ended, they have walls which are very permeable, so that all the interstitial fluid components are able to enter them. The lymphatics thus provide a mechanism for returning to the vascular system those plasma proteins which have escaped from blood capillaries.

Blockage of the lymph drainage (e.g. as a result of carcinoma or other disease) causes the local accumulation of fluid (oedema); this is primarily because of the accumulation of proteins in the interstitial spaces, so that the oncotic pressure difference between blood capillaries and interstitial fluid is reduced, and consequently the reabsorption of tissue fluid into the blood capillaries is impaired.

1.6 Fluid exchanges between the body and the external environment

The body fluid is continually exchanging with the external environment, but the constancy of body weight from day to day indicates that there is equality of fluid intake and output.

1.6.1 Water intake

The following are typical figures for water intake per 24 hours.

Drinking	1500 ml
Food	500 ml
Metabolism	400 ml
Total	2400 ml

The metabolically derived water comes from the oxidation of food substances, e.g. the reaction for glucose oxidation is:

$$C_6H_{12}O_6 + 6O_2 \rightarrow 6CO_2 + 6H_2O$$

1.6.2 Water output

This occurs from the body by several routes; the typical figures per 24 hours are:

Lungs	400 ml
Skin	400 ml
Faeces	100 ml
Urine	1500 ml
Total	2400 ml

The loss from the lungs occurs because air, as it enters the lungs during inspiration, becomes saturated with water vapour as it passes along the respiratory tract. Some of this water is then lost during expiration. The figure of 400 ml is really the minimum loss. In hot dry environments, or in sub-zero temperatures (when the air is very dry because the water vapour has condensed and frozen), the loss of water from the lungs can be considerably greater than 400 ml.

The water loss through the skin (400 ml), termed 'insensible perspiration', occurs at an almost constant rate and reflects evaporative loss from the skin epithelial cells. It is not sweat. Sweating, or 'sensible perspiration', represents an additional loss, of up to 5 l per hour (in order to dissipate excess heat and so maintain constancy of body temperature). The faecal water loss (100 ml) can be greatly increased in diarrhoea, which can lead to the loss of several litres per day.

All of the losses mentioned above (lungs, skin, faeces) can be regarded as disturbances of the body fluid volume. In contrast, the urinary loss, which is normally about 1500 ml/24 h, can be as little as 400 ml, or as much as about 23 l. It is adjusted according to the needs of the body – i.e. the renal fluid output is regulatory (see Chapter 8). However, water intake can be reduced to zero (so that the only water input is that derived from metabolism), whereas the losses from lungs, skin, faeces and kidneys can never be less than about 1200 ml/day. This means that survival without any dietary water is only possible for a short time (generally less than one week).

1.7   Ionic composition of the body fluids

Because the body fluid osmolality is regulated and normally kept constant, it is obvious that the body water content will depend on the overall ionic content of the body, and the distribution of water between the body fluid compartments will depend on the distribution of ions (Figure 1.4). Quantitatively, the most important ions in the body fluids are sodium (the major extracellular ion) and potassium (the major intracellular ion).

1.7.1   Extracellular fluid (ECF)

The major ions of ECF are shown in Figure 1.4. Sodium (Na^+) is the most important cation; chloride (Cl^-) and bicarbonate (HCO_3^-) are the most important anions. The plasma proteins are also anions. In fact the colloids of plasma (i.e. the plasma proteins and lipids) occupy a significant volume (70 ml/litre), so that only 930 ml of each litre of plasma is water, and this can lead to errors in the measurement of other solute concentrations, particularly in pathological conditions involving hyperlipidaemia. Figure 1.4 shows the differences in solute concentrations in total plasma and in the plasma water, and it can be seen that the discrepancy is considerable.

1.7.2   Intracellular fluid

Intracellular fluid composition is shown in Figure 1.4, but it should be noted that it is not the same throughout the body. Different types of cells have

	Plasma		Interstitial fluid	Intracellular fluid		Transcellular fluid		
						CSF	Gastric juice	Ileum secretion
Na⁺	142	(153)	⇌ 145	⇌ 12		141	60	129
K⁺	4	(4.3)	⇌ 4.1	⇌ 150		3	9	11
Cl⁻	103	(109)	⇌ 113	⇌ 4	Exchange regulated by specific transport processes	127	120	116
HCO₃⁻	25	(26)	⇌ 27	⇌ 12		23	0	29
Proteins, g/l	60		0	25		0	v. low	v. low
Osmolality, mosmol/kg	280		280	280		280	280	280
Compartment volume (l)	3.0		12.0	30		Total transcellular fluid volume is estimated to be about 5% of total body water or 2.2l		

Figure 1.4 Ionic concentrations of the body fluids (mmol/l), and the exchanges between body fluid compartments. The figures for plasma are given, in parentheses, as concentrations per litre of plasma water. These values are higher than the concentrations in total plasma, because the ions are present only in the aqueous phase, but 70 ml/l of plasma is protein and lipid. Thus the 930 ml of water per litre of plasma contains sodium at a concentration of 153 mmol/l H_2O, so the concentration of sodium in total plasma is 153 × 930/1000 mmol/l plasma = 142 mmol/l. The compartment volumes indicated are typical values for a 70 kg man. Intracellular fluids in different tissues differ slightly in composition. The values given are for skeletal muscle cells. All values given for intracellular and transcellular fluid are approximate. Note that Na⁺ and K⁺ are actively transported across the cell membrane (although potassium also diffuses readily across this membrane). The protein constituents of the plasma and of intracellular fluid do not diffuse in significant amounts into the interstitial fluid. CSF, cerebrospinal fluid.

different compositions. Furthermore, each cell is compartmentalized and the different cellular sub-compartments have different compositions. Nevertheless, in all cells the most important cation, quantitatively, is K⁺ and the most important anions are the intracellular proteins and phosphate.

1.8 *Ion exchanges between the body and the external environment*

We saw on p. 8 that the *osmolality* of the extracellular fluid depends on the sodium content. Since body fluid osmolality is regulated (see Chapter 7) it follows that the extracellular fluid *volume* also depends on the sodium content of the organism (see also Chapter 8). In a 70 kg man of average build, there are about 4000 mmol sodium, 2500 mmol chloride and 400 mmol bicarbonate.

Potassium is the major ion of the intracellular fluid, and so the osmolality and volume of this body fluid compartment depend primarily on the potassium content. Our 70 kg average man contains about 4000 mmol potassium, of which only about 50 mmol are in the extracellular fluid (and therefore directly accessible to the kidneys).

The diet may contain widely varying amounts of both Na^+ and K^+. The average Western diet contains about 10 g NaCl per day (170 mmol sodium), but a vegetarian diet may contain very little Na^+ and large quantities of K^+. Nevertheless, in spite of this potentially enormous variation in dietary intake, the sodium and potassium contents of the body are normally kept constant. This regulation is brought about mainly by renal control of the tubular reabsorption of the filtered sodium. Much of the rest of this book is concerned directly or indirectly with this process. For potassium, the control is brought about by the tubular reabsorption of about 95% of the filtered K^+ and then the secretion of a variable amount of K^+.

The ability of the body to regulate the plasma K^+ concentration depends to a large extent on movements of K^+ across the cell membrane, between intra- and extracellular fluid. Losses of K^+ from the body (i.e. from extracellular fluid, e.g. in the urine) have only a small effect on the plasma K^+ concentration (denoted $[K^+]$), because K^+ moves out of the cells to maintain the equilibrium between intracellular and extracellular fluid. Similarly, ingestion of large amounts of K^+ has a small effect on plasma $[K^+]$ because the K^+ can enter the cells. Thus the intracellular K^+ tends to 'buffer' changes in extracellular K^+ concentration. Nevertheless, there *are* small changes in plasma $[K^+]$ when total body content of K^+ is altered.

1.9 *Body fluid compartments: methods of measurement*

The volumes of several body fluid compartments can be measured using the **dilution principle**. This principle can be best illustrated by an example.

Suppose one has a bucket containing an unknown volume of water and it is necessary to measure the volume *in situ*, i.e. without emptying the bucket – this is analogous to the measurement of a body fluid compartment. The way to determine the volume is to use a marker substance such as dye. We can take a known quantity, Q mg, of the dye, add it to the water in the bucket, allow it to mix thoroughly so that its concentration is uniform, then take a small sample from the bucket and measure the dye concentration, C (mg/ml), colorimetrically.

$$\text{Concentration of dye } C \text{ (mg/ml)} = \frac{\text{Quantity of dye } Q \text{ (mg)}}{\text{Volume of water (ml)}}$$

$$\text{Volume} = \frac{Q}{C}$$

It will be apparent that the measurement is valid only if the dye is uniformly mixed and if all of it remains within the volume to be measured. In the body, the substances used to measure body fluid compartments may be excreted or metabolized, so we must modify the equation accordingly, to:

Volume of distribution
$$= \frac{(\text{Quantity administered}) - (\text{Quantity metabolized or lost})}{\text{Concentration}}$$

Almost all the substances used in measuring body fluid compartments are excreted. Some are slowly metabolized or incorporated into other body constituents, so the above correction is usually necessary.

The test substance should ideally have the following properties:

(1) not be toxic;
(2) distribute uniformly within the compartment to be measured and not enter other compartments;
(3) not be rapidly metabolized or excreted;
(4) not alter the volume of the compartment being measured.

In practice, two variations of the dilution principle are in common use.

First, there is the single injection method. This is suitable for substances which have a slow rate of removal from the compartment being measured (by renal excretion and/or penetration into other compartments). A known amount of the test substance is injected intravenously, its plasma concentration is determined at intervals and a graph of concentration against time is

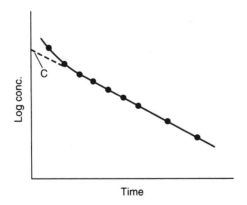

Figure 1.5 Plasma concentration of an injected amount Q of test substance (plotted as log concentration) as a function of time. The straight portion of the graph is extrapolated to zero time to give the plasma concentration, C, which would exist if uniform distribution could occur instantaneously. The volume of distribution is Q/C.

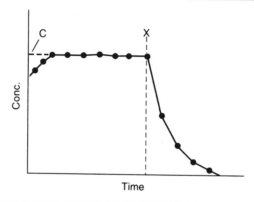

Figure 1.6 Plasma concentration of an infused substance plotted against time. At time *X*, the infusion is stopped and the amount of substance excreted (*Q*) to reduce the plasma concentration to zero is determined. The volume of distribution is then *Q/C*.

plotted (Figure 1.5), using a log scale for the concentration axis. We can then extrapolate the straight portion of the graph back to zero time and this gives us the plasma concentration which would have existed if the substance had instantaneously distributed itself uniformly. Dividing the amount injected by this concentration gives the compartment volume.

The second practical method using the dilution principle is the constant infusion method; this can be used for substances which are rapidly excreted. A loading dose of the test substance is injected to increase its plasma concentration. Then, more of the test substance is infused at a rate which matches the renal excretion rate. This means that when the substance has come to equilibrium, the plasma concentration will remain constant (Figure 1.6). When this occurs, the infusion is stopped and the urine is collected until all of the test substance is excreted. The amount excreted is then the amount which was producing the constant plasma concentration, so we can readily calculate the volume of the compartment:

$$\frac{\text{Amount excreted (mg)}}{\text{Plasma concentration (mg/l)}} = \text{Compartment volume (l)}$$

1.9.1 *Measurement of plasma volume, red cell volume and blood volume*

For the plasma volume measurement, a substance is required which remains confined within the vascular space, i.e. does not cross the capillary endothelium. Plasma proteins are confined in this way, so radio-iodinated human serum albumin can be used or, alternatively, a dye which binds to the plasma albumin. Such a dye is Evans Blue. Since small quantities of albumin escape from the vascular system and small quantities are continually being metabolized, the plasma volume will be slightly overestimated.

When the plasma volume has been determined, the total blood volume can be readily calculated from the plasma volume and the haematocrit. The haematocrit is the percentage of total blood volume made up of red cells. It is obtained by centrifuging a small sample of blood in a closed capillary tube. Suppose the haematocrit is 45%, i.e. plasma is 55% of the blood volume; then

$$\text{Blood volume} = \text{Plasma volume} \times \frac{100}{55}$$

The normal plasma volume is 3 l and the blood volume is about 5 l.

The red cell volume, although it can be calculated from the plasma volume and the haematocrit, can also be measured directly by a dilution method. A small sample of blood is taken and the red cells are incubated in a medium containing radioactive phosphorus (^{32}P) or chromium (^{51}Cr). They are then resuspended (in saline) and reinjected. After a suitable time has elapsed (e.g. 15 min) the dilution of the label is determined. (From the red cell volume and the haematocrit, the plasma volume can be calculated.)

1.9.2   *Extracellular fluid (ECF) volume*

The extracellular fluid volume is very difficult to determine accurately. This is because the ECF is really several compartments – plasma, interstitial fluid and transcellular fluid. The transcellular fluid is a particular problem because, as mentioned on p. 1, it is fluid which is separated from the plasma by another membrane in addition to the capillary endothelium; this additional membrane is normally a layer of cells. Since to measure ECF volume we need a substance which does not enter the intracellular fluid, such a substance will also not penetrate the layer of cells bounding the transcellular fluids.

The substance used to measure ECF volume must be sufficiently diffusible to cross capillary walls rapidly, so that when injected into the plasma it will enter the interstitial fluid; but it must be excluded from the cells – i.e. be unable to cross cell membranes. In fact, because different test substances fulfil these criteria to different extents, the volume of the ECF depends on the substance used to measure it.

The substances which can be used include inulin, mannitol, thiosulphate, radiosulphate, thiocyanate, radiochloride and radiosodium. Inulin (mol. wt 5500) is the largest molecule in the group and, correspondingly, has the smallest volume of distribution (approx. 12 l). It may be excluded from a fraction of the interstitial fluid (e.g. in bone and cartilage). Chloride and sodium ions are not completely excluded from cells and have the largest volume of distribution (18 l). Thiosulphate is probably the most widely accepted substance as a measure of extracellular fluid volume, giving a value of 15 l.

1.9.3   *Interstitial fluid volume*

The interstitial fluid volume cannot be measured directly. It must be calculated as the difference between the extracellular fluid volume (15 l measured with thiosulphate) and the plasma volume (3 l) giving 12 l for the interstitial fluid volume. It will be apparent, however, that the figure will depend on the substance used to measure the extracellular fluid volume.

1.9.4   *Total body water*

Measurement of the total body water is usually accomplished using one of the two isotopes of water (deuterium oxide or tritiated water). With either of these, we find that, in a normal 70 kg average man, about 63% of the body weight is water (i.e. about 45 l). In women, the proportion of the body weight which is water is only about 52%. This difference, as mentioned previously, is due to the greater proportion of fat in women.

1.9.5   *Transcellular fluid volume*

Because the transcellular fluids are separated from the rest of the extracellular fluid by a membrane which is generally composed of intact cells, the marker substances used to measure extracellular fluid volume do not penetrate into the transcellular fluids, or do so extremely slowly. Consequently, transcellular fluid, which is included in the measurement of total body water, is generally excluded from ECF measurements.

$$
\begin{aligned}
\text{Total body water} = &\ \text{extracellular fluid volume} \\
&+ \text{intracellular fluid volume} \\
&+ \text{transcellular fluid volume}
\end{aligned}
$$

We can measure total body water and extracellular fluid volume by the dilution principle:

$$
\begin{aligned}
&\text{Total body water} - \text{extracellular fluid volume} \\
&= \text{intracellular fluid volume} + \text{transcellular fluid volume}
\end{aligned}
$$

Although the turnover of transcellular fluid is large (up to 20 l/day for the gastrointestinal tract), at any moment the absolute volume of transcellular fluid is small and is usually ignored, i.e. it is assumed that

$$
\begin{aligned}
&\text{Total body water} - \text{extracellular fluid volume} \\
&= \text{intracellular fluid volume}
\end{aligned}
$$

Further reading

Acker, B. A. C. van, Kooman, G. C. M. and Arisz, L. (1995) Drawbacks of the constant infusion technique for measurement of renal function. *Am. J. Physiol.* **268**, F543–F552

Manning, R. D. and Guyton, A. C. (1980) Dynamics of fluid distribution between the blood and interstitium during overhydration. *Am. J. Physiol.* **238**, 645–651

Marken Lichtenbelt, W. van, Westerterp, K. R. and Wouters, L. (1994) Deuterium dilution as a method for determining total body water. Effect of test protocol and sampling time. *Brit. J. Nutrition* **72**, 491–497

Michel, C. C. (1988) Capillary permeability and how it may change. *J. Physiol.* **404**, 1–29

Trapp, S. A. and Bell, E. F. (1989) An improved spectrophotometric assay for the estimation of extracellular water volume. *Clin. Chim. Acta.* **181**, 207–212

Walser, M. (1992) Phenomenological analysis of electrolyte homeostasis. In D. W. Seldin and G. Giebisch (eds), *The kidney, physiology and pathophysiology*, 2nd edition, vol. 1, Raven Press, New York, pp. 31–44

Problems

1.1 Is Na^+ an effective or ineffective osmole across

    (a) the capillary wall?
    (b) the cell membrane?

1.2 Is urea an effective or ineffective osmole across

    (a) the capillary wall?
    (b) the cell membrane?

1.3 If 500 ml of a concentrated solution of NaCl (say 500 mosmol/kg H_2O) were infused intravenously, what would happen to

    (a) the intracellular fluid volume?
    (b) the extracellular fluid volume?
    (c) the intracellular osmolality?
    (d) the extracellular osmolality?

1.4 If 500 ml of a concentrated urea solution (say 500 mosmol/kg H_2O) were infused intravenously, what would happen to

    (a) the intracellular fluid volume?
    (b) the extracellular fluid volume?
    (c) the intracellular osmolality?
    (d) the extracellular osmolality?

Essential anatomy of   *2*
the kidney

2.1 *Introduction*

A knowledge of the structure of the kidney is essential to the understanding of its function. In general, structure and function are dealt with together in this book and this chapter therefore consists of only a brief outline of those structural features which are of particular importance, or which are not dealt with in detail elsewhere.

2.2 *General morphology and cellular organization*

The kidneys are situated behind the peritoneum on each side of the vertebral column. In man, the top (upper pole) of each kidney is at the level of the twelfth thoracic vertebra and the bottom (lower pole) is at the level of the third lumbar vertebra. Each kidney is about 12 cm long and weighs about 150 g.

On the medial surface of each kidney (the concave surface) is a slit, the **hilus**, through which pass the renal artery and vein, the lymphatics, the renal nerve and the renal pelvis, which is the funnel-shaped upper end of the **ureter**.

The blood supply to each kidney is usually a single **renal artery** arising from the abdominal aorta. However, there may sometimes be additional small vessels from superior mesenteric, adrenal, spermatic or ovarian arteries.

If a kidney is bisected from top to bottom (Figure 2.1), the cut surface shows two distinct regions, a dark outer region, the cortex, and a paler inner region, the medulla, which is further divided into a number of conical areas, the renal pyramids. The apex of each pyramid extends towards the renal pelvis, forming a papilla. Some animal species have only one pyramid (and papilla) in each kidney. Striations can be seen on the renal pyramids. These are medullary rays, which are attributed to the straight tubular elements (collecting ducts and loops of Henle) and blood vessels (vasa recta) in this region.

Congenital abnormalities of the kidneys and urinary tract

Renal agenesis

Failure to develop kidneys occurs in about 1 in every 2500 foetuses, and is not compatible with life (associated with pulmonary hypoplasia it is known as Potter's syndrome).

Unilateral agenesis

The development of only one kidney affects about 1 in 1000 of the population. The single kidney becomes hypertrophied, and may be abnormally located, and prone to infection and damage. The condition may be associated with other developmental abnormalities.

Hypoplasia

One or both kidneys are small. They are prone to infection and stone formation.

Ectopic kidney

A kidney is in an abnormal location – often in the pelvis. This may cause ureteral obstruction resulting in renal damage and stone formation. The condition affects 1 in 800.

Horseshoe kidney

This affects 1 in 1000. The kidneys are fused across the midline, usually at their lower poles. The horseshoe kidney is generally lower than normal, because the inferior mesenteric artery blocks its ascent. Horseshoe kidneys may be associated with ureteral obstruction and stone formation.

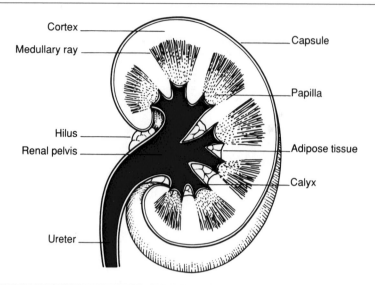

Figure 2.1 A longitudinal section of the kidney to illustrate the main structural features. The components of the kidney are described in the text.

The **renal pelvis** is lined by transitional epithelium and is the expanded upper part of the ureter. Extensions of the pelvis, the calices (singular, calyx) extend towards the papilla of each pyramid and collect the urine draining from it.

The **ureters**, about 30 cm long, are muscular tubes which connect the renal pelvis to the bladder.

2.2.1 *The nephron*

The basic functional unit of the kidney is the nephron. Each human kidney has 1–1.5 million nephrons. The components of the nephron are shown in Figure 2.2.

The nephron is a blind-ended tube, the blind end forming a capsule (Bowman's capsule) around a knot of blood capillaries (the glomerulus). The other parts of the nephron are the proximal tubule, loop of Henle, distal tubule and collecting duct.

The glomeruli, proximal tubules and distal tubules are situated in the cortex, whereas the loops of Henle and the collecting ducts extend down through the medulla. The length of a nephron's loop of Henle depends on the location of the glomerulus, and on this basis we can identify two populations of nephrons in the human kidney (Figure 2.3).

Those nephrons with glomeruli in the outer two-thirds of the cortex are called cortical nephrons and have very short loops of Henle, which only extend a short distance into the medulla (or, indeed, may not reach the

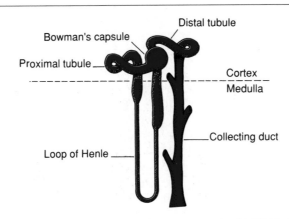

Figure 2.2 A nephron.

medulla at all). In contrast, nephrons whose glomeruli are in the inner one-third of the cortex (juxtamedullary nephrons) have long loops of Henle which pass deeply into the medulla. In man, about 15% of nephrons are long-looped. (The proportion of long-looped nephrons is different in different species, e.g. in the rat the figure is 30%.)

2.2.2 *Glomerulus*

The function of the glomerulus is to produce an ultrafiltrate of plasma, which enters the nephrons (see Chapter 3). In man, the glomerulus has a diameter of about 200 μm. Each glomerulus is supplied with blood by an **afferent arteriole**, which divides within the glomerulus to form the tuft of glomerular capillaries. These capillaries rejoin to form the **efferent arteriole**. Details of glomerular structure are given in Chapter 3.

2.2.3 *Proximal tubules*

The proximal tubule is the first segment of the nephron after the Bowman's capsule; its early part (pars convoluta) is convoluted, but further along it becomes straight (pars recta) and passes down towards the medulla where it becomes the descending limb of the loop of Henle. The length of a human proximal tubule is generally about 15 mm (12–25 mm range) with a diameter (outside) of 70 μm.

The convoluted segment of the proximal tubule consists of cuboidal/columnar cells, which on their luminal surface have a 'brush border'. This consists of

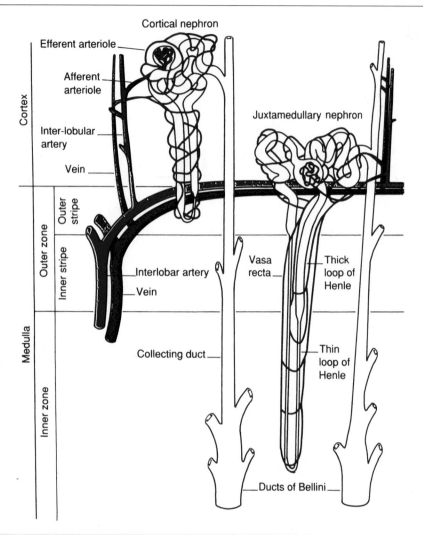

Figure 2.3   Cortical (short-looped) and juxtamedullary (long-looped) nephrons, showing the differences in the blood supply to the two nephron types.

Source: Reproduced with permission from Pitts, R. F. *Physiology of the Kidney and body fluids*, 3rd edn (Year Book Medical Publishers, Chicago, 1974).

millions of microvilli with a density of about 150 per μm^2 cell surface. This brush border increases tremendously the surface area available for the absorption of tubular fluid. Adjacent cells are joined tightly together close to their luminal surface (at a 'tight junction' or zonula occludens), but on the peritubular side of the tight junction there is a space between adjacent cells, the lateral intercellular space (Figure 2.4).

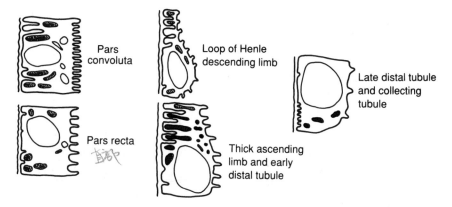

Figure 2.4   The morphology of cells in the various nephron segments. In the diagram, for each cell the tubule is on the right. The cell membrane on this side is the apical, mucosal or luminal membrane. The cell membrane on the opposite side of the cell, resting on the basement membrane (basal lamina), is the basal, serosal or peritubular membrane. In the *pars convoluta* of the proximal tubule, the cells have a dense brush border (microvilli), increasing the luminal membrane surface area by a factor of about 40. Basolaterally, these cells interdigitate with each other extensively. In the *pars recta*, there are fewer mitochondria (suggesting that the transport functions of these cells are less well developed than those of the pars convoluta cells). The brush border is also less extensive (in most species). The cells of the descending limb of the loop of Henle are rather flat, have few interdigitations and sparse microvilli. In the thin ascending limb of the loop of Henle (not illustrated), the cells have a similar flattened appearance, but have dense interdigitations with each other. Generally, the transition from descending-type epithelium to ascending type occurs just before the bend. The transition from the thin ascending limb to the thick ascending limb occurs abruptly at the junction of the inner and outer zones of the medulla (see Figure 2.3). In most species, there are only minor differences between cells of the thick ascending limb and those of the early distal tubule. The late distal tubule and collecting tubule consist of a number of cell types, but the transport functions are essentially similar in both of these anatomical segments.

The cells of the straight part of the proximal tubule, the pars recta (which can also be regarded as the beginning of the loop of Henle), are very similar to those of the convoluted segment, but have a less dense brush border (i.e. fewer microvilli), contain fewer mitochondria and are generally more flattened.

2.2.4   *Loop of Henle and distal tubule*

The loop of Henle is generally considered to begin at the transition from thick-walled tubule to thin-walled tubule (Figure 2.2). The cells of the thin part of the loop of Henle are squamous, i.e. are very thin and flattened. Under the light microscope, the cells of the loop of Henle resemble capillary endothelial cells; however, under the electron microscope, significant differences are apparent. The contours of the cells are extremely complex, the cells interdigitate with their neighbours and there are a few microvilli on the luminal cell surfaces.

The cells of the thin ascending limb of the loop of Henle are structurally similar to those of the descending limb, but there are important functional differences (p. 71), particularly in their permeability properties (and possibly in their capability for active transport). The ascending thin segment of the loop is up to 15 mm long and the external diameter is about 20 μm. The thick ascending segment of the loop of Henle is a cuboidal/columnar epithelium, with cells of a similar size to those in the proximal tubule. However, the cells do not have a brush border like that of the proximal tubule and, although there are many infoldings and projections of the cell on the luminal surface, there are fewer basal infoldings. This portion of the nephron continues into the cortex as the distal tubule. However, the term 'distal' tubule is a purely anatomical description of the location of a portion of the nephron and has no clear functional meaning. Functionally, the thick ascending limb of the loop of Henle meets the cortical collecting ducts.

There are only minor differences between the cells of the thick ascending limb and those of the early distal tubule (Figure 2.4) and the cells of the late distal tubule (also termed the cortical connecting tubule) are almost identical to those of the cortical collecting tubule (or duct). All these cells have infoldings of the basal membrane which, in the early distal tubule and cortical connecting tubule, may surround mitochondria (Figure 2.4). In the cells of the cortical collecting duct (which are sometimes termed 'principal cells'), the infoldings are smaller and rarely have mitochondria between them. These cell types all have Na^+K^+-ATPase activity in their basal membranes.

However, throughout the distal nephron these main cell types are interspersed with cells of a different type, termed intercalated cells. These cells have few or no infoldings of the basal membrane and have little Na^+K^+-ATPase activity. They have many cytoplasmic extensions on the luminal membrane and high carbonic anhydrase activity. In the region where the ascending limb of the loop enters the cortex, it is closely associated with the glomerulus and afferent arteriole, and consists of modified **macula densa** cells (p. 29).

2.2.5   *Collecting duct*

Most of the cells in the collecting duct are cuboidal, with a much less granular cytoplasm than that of the proximal tubule cells and only a few corrugations on the luminal surface. Interspersed with the cells of this type there are a few cells having more granular cytoplasm.

In the cortex, each collecting duct receives about six 'distal tubules' and, as the ducts enter the medulla, they join with each other in successive pairings to form a duct of Bellini up to $200\,\mu$m wide, which drains into a renal calyx. Strictly speaking, anatomically the 'nephron' does not include the collecting duct (and is embryologically distinct from it). However, functionally the collecting duct is an essential part of the nephron unit.

2.3 *Blood supply and vascular structure within the kidney*

The kidneys have a high blood flow, between them receiving a little over 20% of the cardiac output. (This is about five times the flow to exercising skeletal muscle and almost ten times the coronary blood flow, on the basis of flow per unit tissue weight.)

Almost all of the blood which enters the kidneys does so at the renal hilus, via the **renal artery**. The renal artery branches (Figure 2.5) to form several **interlobar arteries**, which themselves branch to give rise to **arcuate** (or arciform) **arteries**, which pass along the boundary between cortex and medulla. From the arcuate arteries, branches travel out at right angles, through the cortex towards the capsule. These are **inter-lobular arteries**, and the **afferent arterioles** which supply the glomerular capillaries branch off from the interlobular arteries.

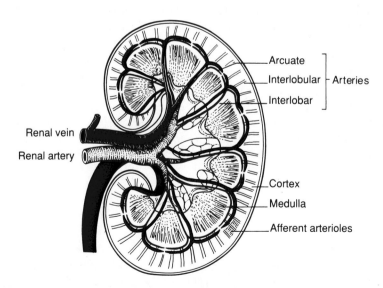

Figure 2.5 A longitudinal section of the kidney, showing the arrangement of the major blood vessels.

The glomerular capillaries are the site of filtration of the blood, the filtrate entering the Bowman's capsule of the nephron. The glomerular capillaries do not drain into a vein; instead they drain into a second arteriole, the **efferent arteriole**. (The efferent arterioles can be regarded as portal blood vessels. Portal vessels carry blood from a capillary network directly to a second capillary network.)

The efferent arterioles from nephrons in the outer two-thirds of the cortex branch to form a dense network of peritubular capillaries, which surround all the cortical tubular elements. The efferent arterioles in the inner one-third of the cortex give rise to some peritubular capillaries, but also give rise to capillaries which have a hairpin course into and out of the medulla, where they are adjacent to the loops of Henle and collecting tubules. These medullary capillaries are *vasa recta* (Figure 2.3). Vasa recta and peritubular capillaries eventually drain into the renal vein which leaves the kidney at the hilus.

More than 90% of the blood which enters the kidney (via the renal artery) supplies the renal cortex, which is perfused at a rate of about 500 ml/min per 100 g tissue (100 times greater than resting muscle blood flow). (The remainder of the renal blood supply goes to the capsule and the renal adipose tissue.) Some of the cortical blood then passes to the medulla; the outer medulla has a blood flow of 100 ml/min per 100 g tissue, and the inner medulla a flow of 20 ml/min per 100 g tissue (man and the rat; in the dog it is only 2.5 ml/min per 100 g tissue).

2.3.1   *Function of the renal blood supply*

In most organs and tissues of the body, the main purpose of the blood supply is to provide oxygen and remove carbon dioxide and other products of metabolism. The renal cortex receives far more oxygen than it requires, so that the arteriovenous O_2 difference is only 1–2%. This is because the high renal blood supply exists to maintain a high glomerular filtration rate. Surprisingly (at first sight), if the renal blood supply is reduced, the arteriovenous O_2 difference does not increase, until the cortical flow is down to about 150 ml/min per 100 g tissue. This is because the blood flow determines the rate of filtration and more than 50% of the O_2 consumption is used for sodium reabsorption (see Chapter 3). So if filtration rate is reduced, reabsorption can occur at a lower rate and O_2 consumption is also reduced. The control of the renal blood supply is considered later (Chapter 7).

In spite of the very high blood flow to the kidneys, the medullary blood supply is no more than adequate for the supply of oxygen to medullary cells, because the vasa recta arrangement causes oxygen to short-circuit the loops of Henle (pp. 78–79).

2.4   *Renal lymphatic drainage*

Renal lymph vessels begin as blind-ended tubes in the renal cortex (close to the corticomedullary junction) and run parallel to the arcuate veins to leave the kidney at the renal hilus. Other lymph vessels travel towards the cortex and may pass through the capsule. The importance of the renal lymphatic drainage is frequently overlooked, but in fact the volume of lymph draining into the renal hilus per minute is about 0.5 ml (i.e. the kidney produces almost as much lymph per minute as urine). Its function is probably to return protein reabsorbed from the tubular fluid to the blood.

2.5   *Juxtaglomerular apparatus*

The ascending limb of the loop of Henle, where it re-enters the cortex and becomes the distal tubule, passes very close to the Bowman's capsule of its own nephron and comes into contact with the afferent and efferent arterioles from its own glomerulus (Figure 2.6). In this area of close association between

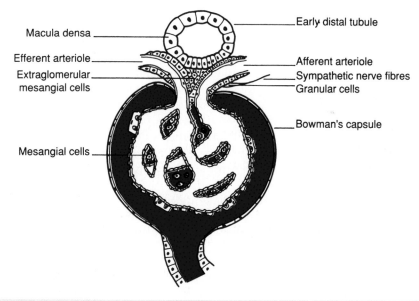

Figure 2.6 The juxtaglomerular apparatus. The beginning of the distal tubule (i.e. where the loop of Henle re-enters the cortex) lies very close to the afferent and efferent arterioles, and the cells of both the afferent arteriole and the tubule show specialization. The cells of the afferent arteriole are thickened, granular (juxtaglomerular) cells and are innervated by sympathetic nerve fibres. The mesangial cells are irregularly shaped and contain filaments of contractile proteins. Identical cells are found just outside the glomerulus, and are termed extraglomerular mesangial cells or 'Goormaghtigh cells'.

the distal tubule and the arterioles is the juxtaglomerular apparatus; it consists of specialized structures in the walls of the afferent arteriole and of the distal tubule. The specialized cells in the distal tubule are **macula densa** cells. They respond to the composition of the fluid within the tubule (p. 112). The specialized cells in the afferent arteriolar wall are **granular cells**. They release the enzyme (hormone) **renin**. The control of the release of renin and its actions are covered in Chapter 9.

The granules in the granular cells contain stored renin and it is likely that the non-granular cells of the juxtaglomerular area are capable of becoming granular cells and of releasing renin. Associated with the juxtaglomerular apparatus are extraglomerular mesangial cells (Goormaghtigh cells, p. 37).

2.5.1   *Erythropoietin*

In addition to their excretory and regulatory function which is the subject of the rest of this book, the kidneys also produce erythropoietin, a glycoprotein hormone of molecular weight 34 000, with 166 amino acids. Small amounts of erythropoietin are produced in the liver, but over 80% of the body's production is from the kidneys. The mesangial cells (Figure 2.6) and the renal tubular cells are the sites of synthesis (both contain erythropoietin messenger RNA).

The stimulus to erythropoietin production and release is hypoxia. Prostaglandin synthesis is thought to mediate the response – i.e. hypoxia stimulates renal cortical prostaglandin synthesis (p. 116), which in turn promotes erythropoietin synthesis. Erythropoietin release is also facilitated by catecholamines acting via β-receptors, presumably on the erythropoietin-producing cells.

The target tissue for erythropoietin is the bone marrow. Erythropoietin-sensitive stem cells in the bone marrow are converted into proerythroblasts and thence to erythrocytes, and this increased erythrocyte production enhances the oxygen-carrying capacity of the blood.

Patients with renal failure often have defective erythropoietin production (and are therefore anaemic) as well as having impaired renal excretory function. Erythropoietin supplementation of such patients is beneficial.

2.5.2   *Bladder and micturition*

Urine is conveyed from the renal pelvis to the bladder, along the ureters, by peristaltic contractions. Dilatation of the pelvis promotes action potentials in the pacemaker cells of the renal pelvis, thus initiating the wave of peristalsis.

The ureters enter the bladder posteriorly, near the base (Figure 2.7). The bladder is composed of two parts, the body, or **fundus**, and the **neck**, which is

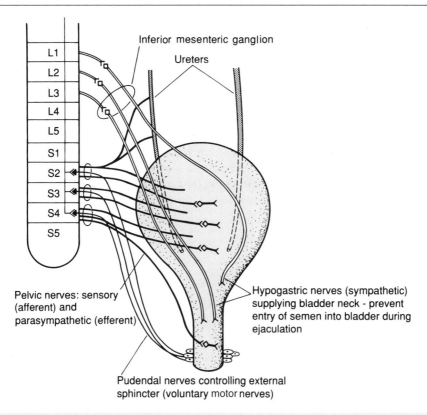

L1
L2
L3
L4
L5
S1
S2
S3
S4
S5

Inferior mesenteric ganglion

Ureters

Pelvic nerves: sensory
(afferent) and
parasympathetic (efferent)

Hypogastric nerves (sympathetic)
supplying bladder neck - prevent
entry of semen into bladder during
ejaculation

Pudendal nerves controlling external
sphincter (voluntary motor nerves)

Figure 2.7 The bladder and its neural control.

also called the **posterior urethra**. In males, there is an additional section of urethra, the **anterior urethra**, which extends through the penis.

The smooth muscle of the bladder, like that of the ureters, is organized in spiral, longitudinal and circular bundles which are not arranged in layers. The bladder muscle is termed the **detrusor muscle** and contractions of this cause micturition (urination). The convergence of the detrusor muscle fibres in the bladder neck constitutes the internal urethral sphincter. Further along the urethra is a sphincter of skeletal (voluntary) muscle, the external urethral sphincter.

The bladder wall has many folds (**rugae**), so that the bladder can expand greatly with little increase in its internal pressure. Micturition, the emptying of the bladder, is an automatic spinal cord reflex, triggered by the filling of the bladder to a critical pressure. However, the reflex can be inhibited or facilitated by higher centres in the brain stem and cerebral cortex.

Urinary incontinence

This is the involuntary loss of urine. It is classified into several types:

1. Stress incontinence

This is loss of urine which occurs accidentally when the intraabdominal pressure is raised (e.g. during a cough). Common causes are weakness of the bladder neck sphincter, laxity of the pelvic floor, or damage to the external sphincter, e.g. as a result of prostatic surgery or during parturition.

2. Urge incontinence

This is involuntary loss of urine occuring immediately the need to void is noticed (i.e. when the bladder contracts). Its causes include urinary tract infection, and excessive reflex contractions of the bladder (which can be caused by urinary tract stones). Urge incontinence is common in the elderly, and can be exacerbated by stroke, Parkinson's disease, and Alzheimer's disease.

3. Overflow incontinence

This occurs when the bladder is full to capacity, and urine then leaks out of the urethra. Causes include outlet obstruction (e.g. an enlarged prostate gland, faecal impaction or urethral stricture), and underactive detrusor muscle.

4. Total incontinence

In this condition there is no voluntary control over voiding. The condition may be congenital, or is caused by neurological damage or disease (e.g. paraplegia, multiple sclerosis).

5. Reflex incontinence

Occasional involuntary voiding

6. *Functional incontinence*

Normal bladder function, but intellectual impairment or immobility prevents the patient from reaching a toilet.

In the elderly, it is common to find that incontinence is due to more than one of the above causes.

Further reading

Caplan, M. J. (1997) Membrane polarity in epithelial cells: protein sorting and establishment of polarized domains. *Am. J. Physiol.* **272**, F425–F429

Glomerular filtration   3

3.1 *The filter*

Urine is initially an ultrafiltrate of the plasma. Ultrafiltration occurs from the glomerulus (a tuft of capillaries) into the Bowman's capsule (the blind end of a nephron) – see Figure 3.1. In moving from the capillary into the Bowman's capsule, the filtrate must traverse three layers. These are:

(1)   the endothelial cell lining of the glomerular capillaries;
(2)   the glomerular basement membrane (non-cellular, composed of connective tissues);
(3)   the visceral epithelial cells of the Bowman's capsule.

Since the properties of the glomerular filter are dependent on these structures, we will begin by looking in more detail at the morphology of the glomerulus and Bowman's capsule.

Figure 3.2 shows the arrangement of the glomerular components. The endothelial cells which form the glomerular capillaries have thin, flattened cytoplasmic areas, but the nuclei are large, may be folded or distorted, and bulge out into the capillary lumen. There are few mitochondria in these cells. Adjacent endothelial cells are in contact with each other, and the cells have many circular fenestrations (pores), with a diameter of about 60 nm (Figure 3.3).

Immediately beneath the endothelial cells is a basement membrane; it forms a continuous layer, and is thought to be the main barrier to the filtration of large molecules. It consists of glycoproteins, mainly type IV collagen, laminin, and fibronectin, all of which are negatively charged.

The third layer of the filter consists of the visceral epithelial cells of Bowman's capsule. These cells are called podocytes and have an extremely complex morphology. The cell body has projections from it (trabeculae) which encircle the basement membrane around the capillary. From the trabeculae, many smaller processes (pedicels) project. The pedicels interdigitate with those of adjacent trabeculae and it may be that, in living animals, the pedicels are in contact with each other (although after fixation for microscopy, gaps appear to be present); this contact means that substances passing through the slits (slit pores)

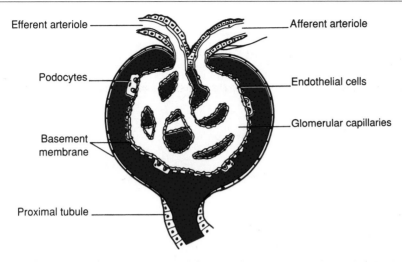

Figure 3.1 Diagrammatic representation of the glomerulus and Bowman's capsule. Bowman's capsule is the blind end of the nephron and is invaginated by the capillary loops of the glomerulus. The figure is diagrammatic rather than anatomically accurate, in several respects. In reality, the afferent arteriole, as it enters Bowman's capsule, branches to form about six vessels. Each of these subdivides to form a knot of about 40 glomerular capillary loops, and there are many interconnections between these capillary loops. The epithelium of Bowman's capsule in contact with the capillaries is a highly specialized cell layer – the podocytes. This is shown in the diagram as covering only a small area of the capillaries, whereas in reality it covers the entire surface of the glomerular blood vessels. A way of envisaging this is to think of Bowman's capsule as a balloon. The capillaries pressed into the 'balloon' will be covered completely by it. The basement membrane of the outer part of Bowman's capsule is continuous with the basement membrane of the rest of the nephron and is also present between the capillary endothelium and the podocytes (see Figure 3.2).

between adjacent pedicels must pass close to the surface coating of the pedicels, and this could influence the filtration behaviour of large, charged molecules (see below). Although, at various times, special filtration functions have been attributed to the podocytes, it seems likely that the main function of these cells is to lay down and maintain the basement membrane.

3.1.1 *Functions of the components of the glomerular filter*

The current view of the three components of the glomerular filter is as follows:

(1) The capillary endothelium acts as a screen to prevent blood cells and platelets from coming into contact with the main filter, which is:
(2) The basement membrane, which allows passage of molecules depending on their molecular size, shape and charge, and is the main filtration barrier.

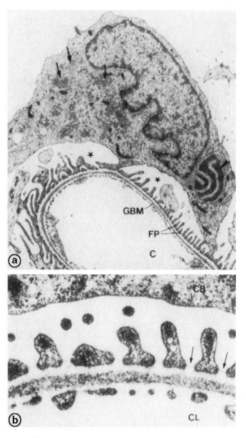

Figure 3.2 The arrangement of the elements of the glomerulus through which the fil-trate must pass to enter the nephron. a) Electron micrograph (×5000 approx) of a podocyte and the glomerular filter of the rat. The podocyte contains well developed Golgi apparatus (arrows); processes (the pedicels or foot processes, FP) from the podocyte rest on the glomerular basement membrane, GBM. The capillary lumen C is also shown. b) Electron micrograph (×50000 approx) showing more details of the glomerular filter. The gaps (filtration slits) between foot processes are arrowed. CB is the cell body of a podocyte, and CL is the capillary lumen. Note the fenestrae in the capillary endothelium (see also Figure 3.4). (From Kriz and Kaissling (1992), repro-duced with permission.)

(3)   The visceral epithelial layer of Bowman's capsule (the podocytes) main-tains the basement membrane, may phagocytose macromolecules (as do mesangial cells) and may have a role in permselectivity which supplements that of the basement membrane.

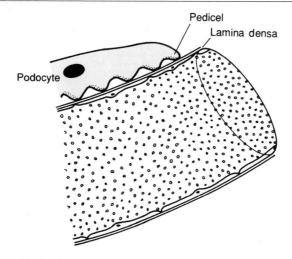

Figure 3.2 (Continued) c) Diagrammatic picture showing glomerular capillary endothelium, basement membrane (lamina densa) and the podocytes.

3.1.2 *Mesangial cells*

In the central part of the glomerular tuft there are a number of irregularly shaped cells, termed mesangial cells. These are actively phagocytic and may prevent the accumulation in the basement membrane of macromolecules which have escaped from the capillaries. The cells may also have a structural role in holding the delicate glomerular structure in position and, in addition, are capable of contraction (i.e. behave like smooth muscle cells) and so may be able to modify the surface area of the glomerular capillaries available for filtration.

Outside the glomerular area, close to the macula densa, are cells identical to mesangial cells. These are termed extraglomerular mesangial cells, or 'Goormaghtigh cells' (Figure 2.6).

3.2 *Glomerular filtration process*

An almost protein-free ultrafiltrate passes into Bowman's capsule from the glomerular capillaries. Molecular size is the main determinant of whether a substance will be filtered or will be retained in the capillaries. However, molecular shape and charge also influence the filtration process, although these factors are of significance only for large molecules. For example, the rate of filtration of albumin (mol. wt 69 000), which has a negative charge, is only about 1/20 that of uncharged dextran molecules of the same molecular weight

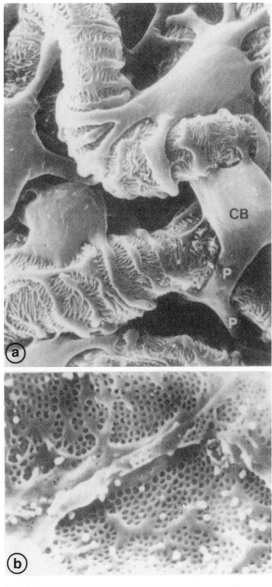

Figure 3.3   a) Scanning electron micrograph (× 3000 approx) of the outer surface of the glomerular capillaries of the rat kidney. CB is the cell body of a podocyte, and the large processes P coming from this are trabeculae, which wrap around the glomerular capillaries. From the trabeculae extend many hundreds of smaller processes, the foot processes or pedicels. Note that the interdigitating foot processes arise from different trabeculae. b) Inner surface of a glomerular capillary (rat, × 16 000) showing the numerous holes (fenestrae) in the capillary endothelium. (From Kriz and Kaissling (1992), reproduced with permission.)

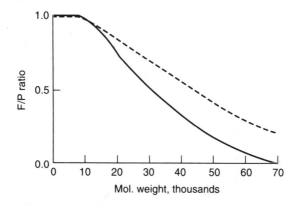

Figure 3.4 The filtration properties of the glomerular filter. The solid line shows the filtrate/plasma concentration ratio (*F/P* ratio) for substances which occur naturally in the body. For substances up to molecular weight 7000, the filter permits free filtration, i.e. solutes pass through as readily as the solvent (H_2O), and the *F/P* ratio is 1. Higher molecular weight molecules (proteins) are retarded by the filter, so the *F/P* ratio is less than 1. Filtration of molecules with a molecular weight of 70 000 and above is insignificant. Molecular charge, as well as molecular size, determines the filtration of large molecules. The dotted line shows the *F/P* ratio of uncharged dextran molecules. They are filtered to a greater extent than are (negatively charged) protein molecules of the same molecular weight.

(Figure 3.4). This finding suggests that the glomerular filtration barrier (the basement membrane and the pedicels) has fixed anions, which repel anionic macromolecules and thereby hinder or prevent the filtration of such molecules.

What exactly is meant by the term 'filtration' (or ultrafiltration)? Filtration is the bulk flow of solvent through a filter, carrying with it those solutes which are small enough to pass through the filter. The prefix 'ultra' simply means that the filter operates at the molecular level (in contrast to the macroscopic particle level of conventional filters).

In the glomerulus, the molecular-weight cut-off for the filter is about 70 000. Plasma albumin, with a molecular weight of 69 000, passes through the filter in minute quantities (retarded also by its charge, as mentioned above). Smaller molecules pass through the filter more easily, but the filter is freely permeable only to those molecules with a molecular weight less than about 7000. The relationship between molecular weight and filtration characteristics is shown in Figure 3.4.

Since the glomerular filter permits the free passage of molecules of molecular weight less than 7000, the initial glomerular filtrate will contain small molecules and ions (e.g. glucose, amino acids, urea, sodium, potassium) in almost exactly the same concentrations as the afferent arteriolar concentrations, and similarly the efferent arteriolar concentrations of such substances

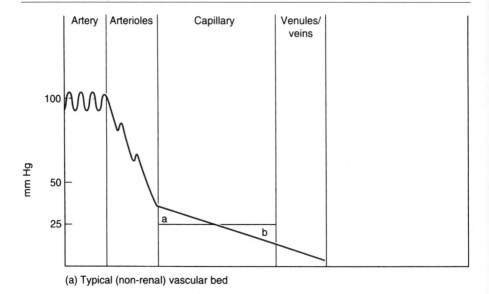

(a) Typical (non-renal) vascular bed

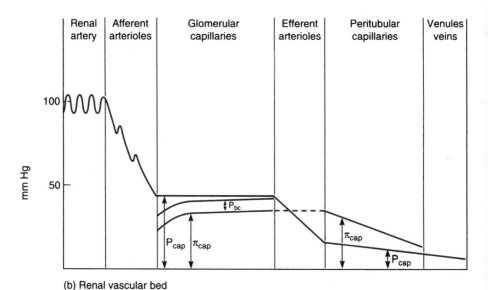

(b) Renal vascular bed

Figure 3.5 The fall in mean blood pressure from arteries to capillaries. (a) Typical systemic vascular bed (e.g. in muscle). The main resistance to flow, and hence the biggest pressure drop, is in the arterioles. The plasma protein osmotic pressure (oncotic pressure) in the capillaries (Π_{cap}) is normally 25 mmHg. In region *a* of the capillaries, the hydrostatic pressure exceeds the plasma protein osmotic pressure and there is net fluid movement out of the capillary. In region *b*, the pressure gradient is reversed, and there is net fluid movement into the capillary. (Only the capillaries have

will not have been significantly altered by the filtration process. (There will be very small concentration differences due to the Gibbs–Donnan effect across the glomerular capillary membrane.)

In man, the glomerular filtration rate (GFR) is about 180 l/day (125 ml/min). Since the filtrate is derived from plasma, and the average person has only 3 litres of plasma, it follows that this same plasma is filtered (and reabsorbed in the tubules) many times in the course of a day. Yet, despite the magnitude of glomerular filtration, the forces determining filtration are not fundamentally different in nature from those forces determining the formation of tissue fluid through capillaries elsewhere in the body, and it is of interest to consider the process of tissue fluid formation and reabsorption in ordinary capillaries, before considering the special case of the glomerular capillaries.

Vascular beds, in any tissue, consist of a set of different kinds of vessels, connected in series. In typical systemic vascular beds, e.g. in skin or muscle, the vessels are: arteries, arterioles, capillaries, venules and veins.

The pressure drop along an idealized capillary from the arteriolar end to the venous end is shown in Figure 3.5a. At the arteriolar end, the forces causing fluid to leave the capillary and enter the interstitial fluid are greater than the forces tending to retain fluid in the capillary and so tissue fluid formation occurs. In contrast, at the venous end of the capillary, the forces leading to the re-entry of fluid into the capillaries exceed those forcing fluid out, and tissue fluid reabsorption occurs. Consequently, there is a balance between formation and reabsorption of tissue fluid.

Figure 3.5 (Continued) highly permeable walls, and hence it is only in the capillaries that the plasma–interstitial fluid pressure gradients determine fluid movements.) (b) Renal blood vessels. By comparison with (a), it can be seen that the pressure falls less along the renal afferent arterioles than in muscle arterioles (to 45 mmHg instead of 32 mmHg) and the pressure in the glomerular capillaries (P_{cap}) is maintained and falls very little along the length of the capillary. This is because of the resistance vessel beyond the capillary, the efferent arteriole. The pressure in the glomerular capillaries, P_{cap} (45 mmHg), is a force for the formation of glomerular filtrate. This force is opposed by the hydrostatic pressure in the Bowman's capsule (P_{bc}), normally 10 mmHg, and by the plasma protein osmotic pressure (oncotic pressure) in the capillaries, Π_{cap}, which is initially 25 mmHg but increases as filtration proceeds and the proteins in the glomerular capillaries become more concentrated. Eventually, the forces opposing filtration ($\Pi_{cap} + P_{bc}$) can become equal to the forces favouring filtration so that net filtration ceases, i.e. filtration equilibrium is achieved. This has been shown to occur in the Munich-Wistar rat strain. However, in the dog, and probably in man, filtration equilibrium is not quite achieved. There is a further fall in hydrostatic pressure in the efferent arterioles, so that hydrostatic pressure, P_{cap}, in the peritubular capillaries is very low, whereas the oncotic pressure, Π_{cap}, in these capillaries is close to 35 mmHg because the plasma proteins were concentrated in the glomerular capillaries. Fluid therefore is taken up into the peritubular capillaries (i.e. these capillaries reabsorb fluid) so that the oncotic pressure of the proteins in the peritubular capillaries is progressively reduced by this fluid uptake.

In the glomerular capillaries, the anatomical arrangement of the blood vessels alters the magnitude of the forces causing fluid movements across the capillaries, as shown in Figure 3.5b. The sequence of blood vessels is: interlobular arteries, afferent arterioles, glomerular capillaries, efferent arterioles, peritubular capillaries, venules and veins. The presence of a second resistance vessel, the efferent arteriole, means that the hydrostatic pressure in the glomerular capillaries falls very little throughout their length. This pressure is slightly higher, at 45 mmHg, than that in the capillaries of most other vascular beds (about 32 mmHg). This does not necessarily mean that filtration out of the glomerular capillaries occurs along their entire length, because, as the filtration process occurs, the non-filterable substances (the plasma proteins) become progressively more concentrated. Thus the oncotic pressure in the capillaries increases and eventually, if the oncotic pressure reaches about 35 mmHg, the net ultrafiltration pressure has been reduced to zero, and there is filtration pressure equilibrium.

Current evidence indicates that filtration pressure equilibrium is achieved in some rat strains, but not in dogs, in which filtration will continue along the whole length of the glomerular capillaries. It is likely that the filter in man behaves as in Figure 3.5, i.e. filtration pressure equilibrium is not achieved. It should be noted that, if filtration pressure equilibrium is attained, alterations of the filtration coefficient (K_f, see below), would need to be very large (decreases) to affect the GFR, but small changes in K_f will have a considerable effect on GFR if there is filtration pressure disequilibrium.

The forces governing glomerular filtration are called Starling's forces, i.e. hydrostatic pressure gradients and oncotic pressure gradients. Putting this into a mathematical formulation, we can say that:

$$\text{GFR} \propto \text{Forces favouring filtration} - \text{forces opposing filtration}$$

$$\propto (P_{cap} + \Pi_{bc}) - (P_{bc} + \Pi_{cap})$$

where P_{cap} is glomerular capillary hydrostatic pressure normally 45 mmHg, Π_{bc} is the oncotic pressure in Bowman's capsule, P_{bc} is the hydrostatic pressure in Bowman's capsule and Π_{cap} is the oncotic pressure in the glomerular capillaries. Since negligible amounts of protein enter the Bowman's capsule, Π_{bc} is normally zero, and

$$\text{GFR} \propto P_{cap} - P_{bc} - \Pi_{cap}$$

P_{bc} is usually about 10 mmHg, and Π_{cap} changes as filtration proceeds, as described above.

3.2.1   *Glomerular capillary permeability*

In order to convert the above equation for GFR into real measurements, we have to introduce the term K_f, the filtration coefficient, which is the product of glomerular capillary permeability and the area of capillary available for filtration, i.e.

$$\text{GFR} = K_f(P_{cap} - P_{bc} - \Pi_{cap})$$

The permeability of glomerular capillaries is about 100 times greater than the permeability of capillaries elsewhere in the body.

3.3 *Composition of the glomerular filtrate*

Table 3.1 shows the plasma and Bowman's capsular concentrations of some major solutes. The normal urinary concentrations are also shown. It can be seen that, whereas the filtration process produces no significant changes in the concentrations of small solutes, the tubular fluid has been considerably modified by the time it is excreted. These modifications to the filtrate, which occur in the rest of the nephron, will be considered in the next two chapters.

Although the quantity of protein filtered at the glomerulus is small, its loss in the urine would represent a considerable wastage over the course of a day. Most of the filtered protein is reabsorbed in the proximal tubule and enters the renal lymph vessels. The amount of protein entering the renal lymph vessels per day is about 30 g, which is equal to the amount of protein entering the glomerular filtrate, i.e. essentially all of the filtered protein is reabsorbed into the renal lymph vessels.

Table 3.1 Normal solute concentrations in plasma, plasma water, the initial ultrafiltrate in Bowman's capsule and the final urine. The differences in solute concentrations between Bowman's capsule fluid and the plasma water are caused by Gibbs–Donnan effects (see Chapter 1). The figures for the urinary concentrations of electrolytes are typical values; wider variations are possible

	Plasma (mmol/l)	Plasma water (mmol/l)	Bowman's capsule (mmol/l)	Urine (mmol/l)
Sodium	142	153	142	50–150
Potassium	4.0	4.3	4.0	20–100
Chloride	103	109	113	50–150
Bicarbonate	24–27	26–29	27–30	0–25
Glucose	5.5	5.9	5.9	0
Protein	6 g/100 ml	—	0.020 g/100 ml	<0.010 g/100 ml

3.4 Filtration fraction

The renal blood flow is large in relation to the size of the kidneys (about 1.1 l/min), but only a relatively small fraction of this is filtered. In fact, since a proportion of the blood is cells (which are not filterable), the **renal plasma flow** is the amount of fluid entering the kidney which is potentially filterable. The renal plasma flow is about 600 ml/min.

The normal glomerular filtration rate is 125 ml/min. Thus the **filtration fraction** is 125/600 or approximately 20%, i.e. of every 600 ml of plasma arriving at the glomeruli, about 475 ml continues into the efferent arterioles.

3.4.1 *Glomerular filtration pressure as a force for tubular flow*

The net filtration pressure must be sufficient to move plasma out of the glomerular capillary and into the Bowman's capsule, but in addition it must also maintain the flow of fluid along the nephron. The pressure in the Bowman's capsule (P_{bc}) must therefore be high enough to overcome the viscosity of the tubular fluid and its friction against the tubule walls, and to maintain the tubules in a patent form against the renal interstitial pressure tending to compress the tubules. Occlusion of the renal blood supply, so that filtration ceases, causes the collapse of the tubular lumens.

3.5 Tubulo-glomerular feedback

In addition to being able to determine the overall glomerular filtration rate (GFR; Chapter 7), it is possible to determine the filtration rate of single nephrons; this single nephron GFR (SNGFR) is greater in juxtamedullary nephrons than in cortical nephrons (50 nl/min compared with 30 nl/min).

There is evidence that, for individual nephrons, the SNGFR is determined or influenced by the composition of the tubular fluid in the distal nephron. (This composition will be influenced by the filtration rate.) The mechanism of this tubulo-glomerular feedback can be divided into three components:

(1) A luminal component, whereby some characteristic of the tubular fluid is recognized by the tubular epithelium.
(2) A mechanism whereby the signal is transmitted to the glomerulus.
(3) An effector mechanism to adjust the rate of glomerular filtration.

There is uncertainty about the precise details of all three of these components. The luminal sensor is thought to be the macula densa, but it is not clear precisely which variable is sensed. It does not appear to be the distal tubular sodium chloride concentration *per se*, but could be the osmolality, the delivery of osmoles or the rate at which sodium chloride (or Na^+ or Cl^-) is transported into the macula densa cells (which will depend not only on the NaCl concentration but also on the tubular flow rate). A change in NaCl transport across the macula densa cells could directly alter the NaCl concentration and the osmolality in the juxtaglomerular interstitium, but whether this is the signal mechanism or whether it is angiotensin II or some other humoral agent, remains in doubt, as does the way in which the SNGFR is adjusted.

The purpose of such a mechanism can be regarded as preventing the overloading of the reabsorptive capacity of individual nephrons, i.e. if the tubular NaCl load is too high, filtration by that nephron is decreased.

Glomerular diseases and abnormalities

Hereditary conditions affecting the glomeruli

1. *Alport's Syndrome*
This is an X-linked condition in which the symptoms mainly affect males. It affects about 1 in 5000 of the population and accounts for about 3% of all cases of chronic renal failure. It is essentially a connective tissue disorder with abnormal type IV collagen in the glomerular basement membrane. Sufferers also have haematuria, sensorineural deafness, and eye abnormalities.

2. *Fabry's Syndrome*
This is an X-linked disorder in which a complex lipid (glycosphingolipid) is deposited in the kidneys, as well as in the skin and blood vessels. It is a metabolic defect arising from the deficiency of an enzyme, galactosidase A.

Further reading

Comper, W. D. and Glasgow, E. F. (1995) Charge selectivity in kidney ultrafiltration. *Kidney Int.* **47**, 1242–1251

Dworkin, L. D. and Brenner, B. M. (1992) Biophysical basis of glomerular filtration. In D. W. Seldin and G. Giebisch (eds), *The kidney: physiology and pathophysiology*, 2nd edition, Raven Press, New York, pp. 979–1016

Kriz, W. and Kaissling, B. (1992) Structural organization of the mammalian kidney. In: *The kidney: physiology and pathology* (eds. D. W. Seldin and C. Giebisch), Raven Press, New York

Problems

3.1   If there is a small decrease in the glomerular filtration coefficient, K_f, what will happen to the efferent arteriolar oncotic pressure if

(a)   there is filtration equilibrium?
(b)   there is filtration disequilibrium?

3.2   Why is the glomerular capillary hydrostatic pressure maintained along the length of the capillaries, instead of decreasing (as in capillaries in other parts of the body)?

3.3   Of the plasma delivered to the glomerular capillaries, only 20% is filtered into the Bowman's capsule. Wouldn't it be more efficient to filter all of it?

Tubular transport　　　4

4.1 *Introduction*

In the previous chapter (p. 39) the term 'filtration' was defined, in the context of glomerular filtration. The fluid entering the nephron by glomerular filtration has a composition very similar to that of plasma, except that plasma proteins, which cannot readily cross the glomerular filtration barrier, are almost completely absent from the tubular fluid. However, the final urine which leaves the nephrons to enter the bladder and be excreted is very different from the initial glomerular filtrate, because the composition of the filtrate is modified by reabsorption and secretion of specific substances – i.e. there are selective reabsorption and secretion processes (Figure 4.1).

Transport of solutes (and water) across the renal tubular epithelium can occur *between* the cells (paracellular movement) or *through* the cells (transcellular movement), as shown in Figure 4.2. A substance transported through the cells has to cross two cell membranes (the apical and basal membranes of the cell).

In the kidney tubules, the terms 'reabsorption' and 'secretion' indicate the direction of movement; **reabsorption** is movement of a substance from the tubular fluid, into or between the tubule cells, and thence into the blood, whereas **secretion** is movement from the blood, through or between the tubule cells, into the tubular fluid. Neither term conveys any information about the nature of the forces causing the movement. In Chapter 1, we saw that active transport is important in enabling cells to maintain their normal ionic composition and volume. In the kidney, active transport processes not only serve this purpose but also make possible the movement of substances right across the tubular epithelium.

4.2 *Primary and secondary active transport*

Conventionally, the transport of solutes has been regarded as 'active' if metabolic energy is required and 'passive' if metabolic energy is not required. This

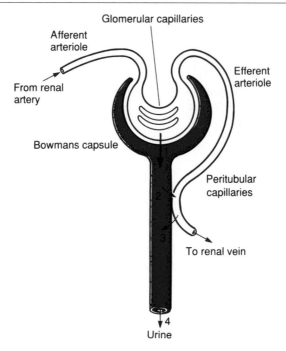

Glomerular capillaries

Afferent
arteriole

Efferent
arteriole

From renal
artery

Bowmans capsule

Peritubular
capillaries

2

3

To renal vein

4
Urine

Figure 4.1 Diagrammatic nephron, to illustrate the terminology of renal function. 1. *Filtration*, which occurs at the glomeruli. 2. *Reabsorption* and 3. *Secretion*, which occur along the nephron. 4. The fluid which leaves the nephrons to be conveyed to the bladder is *excreted*.

distinction is too simplistic, since many transport processes which do not directly require metabolic energy would nevertheless not occur if metabolic energy were not available.

Genuine 'active transport' means the direct coupling of ATP hydrolysis to a transport process in the cell membrane, and can be termed **primary active transport**. The most important primary active process in the nephron is the Na^+K^+-ATPase, located on the basal (and basolateral) side of the cells lining the nephron. The Na^+K^+-ATPase accounts for much of the oxygen consumption of the kidney (28 μmol Na^+ is reabsorbed per μmol oxygen consumed) and enables the nephrons to reabsorb over 99% of the filtered sodium.

Other primary active transport mechanisms in the nephron include a Ca^{2+}-ATPase, an H^+ (proton)-ATPase (Figure 4.3), and an H^+K^+-ATPase.

The ATPases, and particularly Na^+K^+-ATPase, establish ionic gradients across the nephron cell membranes and these gradients then act as driving forces for the reabsorption or secretion of many other solutes, which are thus transported by 'secondary active transport', or 'cotransport'. Although the transport of these other solutes is not directly linked to ATP breakdown, nevertheless if primary active transport (ATP breakdown) were not occurring,

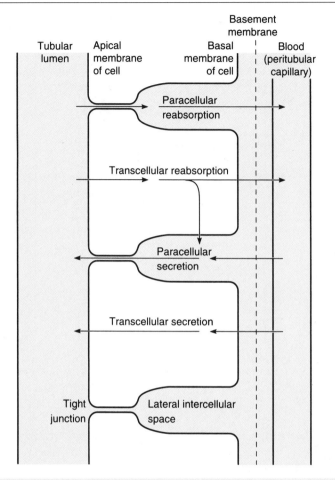

Figure 4.2 How substances can cross the renal tubular epithelium, which consists of a single layer of cells. At the luminal side, adjacent cells are in contact (the tight junction), whereas towards the basal side of the cells, there are gaps between adjacent cells (lateral intercellular spaces).

the transport of the other solutes would not occur. The sodium gradient is the most important driving force for secondary active transport. When the solute which is being transported moves with the Na^+ gradient, the process is termed symport (or, confusingly, cotransport); where the solute which is being transported moves in the opposite direction to the Na^+ gradient (but nevertheless is dependent on Na^+ moving down the gradient), the process is termed antiport (or countertransport). More details are presented in relation to specific solutes, in later chapters. Antiport and symport occur via specific protein carrier molecules (in the cell membranes), termed transporters.

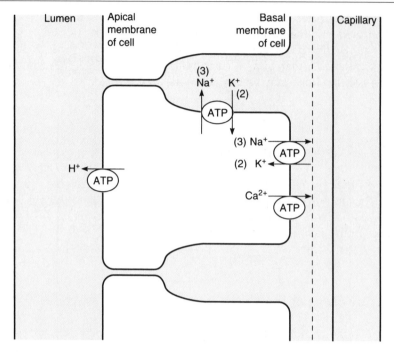

Figure 4.3 The major primary active transport processes in the nephron.

4.3 *Ion channels, uniporters and facilitated diffusion*

In addition to ATPases and transporter molecules, epithelial cell membranes also contain proteins which constitute ion 'channels'. There is a sodium ion channel in the apical membrane of the cells throughout the nephron. This channel is closed by the drug amiloride, and opened by a number of hormones (see e.g. p. 116). There are also Cl^- channels and K^+ channels in the apical membrane in all segments of the nephron. Ion channels allow much faster rates of transport than either ATPases or transporter molecules, but the cell membranes contain relatively few ion channels. A typical nephron cell (e.g. in the proximal tubule) may contain only 100 Na^+ channels (and a similar number of chloride channels) in the apical membrane, but each of these can conduct 10^6–10^8 ions per second. In contrast, there may be 10^7 Na^+ K^+-ATPase molecules in the basal membrane of the same cell, each pumping about 100 Na^+ ions per second.

Cell membranes also contain a number of protein molecules which transport a single substance across the membrane. These are termed **uniporters**, and are driven by the concentration gradient for the subtance concerned. Examples of this kind of transporter include the exit of glucose from proximal tubule cells across the basolateral membrane, and urea transport in the collecting tubule.

Transport which occurs via channels or uniporters, and which therefore occurs at a faster rate than simple diffusion across the lipid bilayer of the cell membrane would allow, may be termed **facilitated diffusion**.

4.4 *Paracellular movement*

This is movement of substances through the spaces between the cells of the nephron. Driving forces for paracellular movement are concentration, osmotic or electrical gradients. Such movements are discussed, in relation to specific substances, in the following chapters.

4.5 *Water absorption*

There is no active water reabsorption. Water reabsorption along the nephron osmotically follows solute absorption and can be both transcellular and para-cellular. The water which passes transcellularly has to cross the apical and basal membranes of the nephron cells, and until very recently the rate at which water can cross many cell membranes was puzzling. For example, water movements across the renal proximal tubule cells (and across the membranes of many non-renal cells, such as red blood cells), seemed to occur at rates much faster than would be possible by diffusion through the lipid bilayer cell membrane. The puzzle was resolved with the discovery of specific water channels, called **aquaporins**, in cell membranes. Five different aquaporins have now been identified. Aquaporin-1 is widely distributed in the body, e.g. in lung and brain cells, and in the kidney, aquaporin-1 is responsible for the high water permeability of both the apical and the basolateral membranes of the proximal tubule cells. Aquaporin-2 is present in the collecting tubules, and the incorporation of this aquaporin into the collecting tubule cell apical membranes is controlled by antidiuretic hormone (see Chapter 6). Aquaporin-3 is in the basolateral membranes of the collecting tubule cells and allows water reabsorbed across the apical membrane via aquaporin-2 to leave the cells and enter the interstitium. Aquaporins 4 and 5 have primarily non-renal distribution (in brain, lungs and salivary glands). Details of water movements and the roles of aquaporins are presented in the following chapters in relation to specific nephron segments.

The water which moves across the tight junctions (i.e. paracellularly) by osmosis can also carry some solutes with it, a process termed **solvent drag**.

4.6 *Fractional excretion and fractional reabsorption*

These terms are used to denote the excretion or reabsorption of solutes or water as a percentage of the amount filtered. Thus if we were to say that the fractional reabsorption of sodium in the proximal tubule was 70%, this would mean that 70% of the amount of sodium filtered was reabsorbed at this site. A comparison of the fractional reabsorption of a substance with the fractional reabsorption of water enables us to determine whether the concentration of the substance has increased or decreased along the nephron. For example, in the proximal tubule, the fractional reabsorption of water and sodium are essentially the same, i.e. there is no change in the sodium *concentration* (although the *amount* of both sodium and water in the tubule is reduced). However, the fractional reabsorption of glucose is almost 100%, so the glucose concentration in the tubular fluid falls. In contrast the fractional reabsorption of magnesium in the proximal tubule is only about 30%, so the magnesium concentration in the proximal tubular fluid increases, in spite of magnesium being reabsorbed at this site, simply because relatively more water is reabsorbed.

Further reading

King, L. S. and Agre, P. (1996) Pathophysiology of the aquaporin water channels. *Annu. Rev. Physiol.* **58**, 619–648

The proximal tubule   5

5.1 *Morphology of proximal tubule cells*

The morphology of the proximal tubule cells was covered briefly in Chapter 2, but will be considered in more detail here. The proximal tubule is divisible into the convoluted portion, or pars convoluta, which begins immediately behind the glomerulus, and the straight portion, or pars recta, which passes into the medulla to become the loop of Henle. The cells of these two portions have somewhat different structures (Figure 2.4) and there are cells of an intermediate type linking the two portions. The transport functions of the proximal tubule are primarily dependent on the pars convoluta cells.

Adjacent proximal tubule cells are in contact with each other at the luminal side (the tight junction), but there are gaps between the cells – lateral intercellular spaces – at the peritubular side. The luminal surface has a brush border of microvilli, which greatly increases the surface area available for absorption (Figure 2.4).

There are some differences in the transport properties of the early proximal tubule (pars convoluta), compared with later parts (pars convoluta and pars recta), particularly in relation to Cl^- transport (see below).

5.2 *Proximal tubular handling of sodium*

The reabsorption of sodium is of great significance, not only because of the importance of sodium to the body, but also because the reabsorption processes for many other substances (including chloride, water, glucose and amino acids) are dependent on sodium reabsorption.

The proximal tubule is highly permeable to sodium in both directions; net absorption from the tubule is the result of slightly greater efflux from the lumen than influx into it and, in fact, net sodium entry into the peritubular capillaries is only about 20% of the unidirectional sodium efflux from the tubular lumen, because there is a high backflux; some of this backflux is via intercellular channels, i.e. between the cells (paracellular movement).

5.2.1   *Sodium entry*

The proximal tubule cells have a negative intracellular potential of approximately $-70\,mV$ relative to both luminal fluid and peritubular fluid, and the cells also have a low intracellular sodium concentration (less than $30\,mM$). So sodium movement from the luminal fluid into the cell is down a large electrical gradient and also down a chemical concentration gradient; thus sodium entry into the cell occurs passively. However, this entry of sodium is carrier-mediated. Many other solutes are transported by Na^+-linked symport and antiport. Most (80%) of the sodium entering the tubule cells does so in exchange for H^+ secretion, which in turn leads to the entry into the cells of both Cl^- and HCO_3^- (as CO_2; Figures 5.1 and 5.6).

5.2.2   *Sodium extrusion*

The Na^+K^+-ATPase extrudes sodium from the cells against electrical and chemical gradients. This is accompanied by the entry into the cells of potassium ions. However, this has little effect on the intracellular K^+ concentration, since K^+ can readily cross cell membranes and so rapidly diffuses out of the cells. The ratio of transport is not $1:1$. In fact, $3Na^+$ leave for $2K^+$ entering.

   The active extrusion of sodium from the tubule cells occurs almost entirely across the basolateral and basal surfaces of the cells, and much of this transport is directed into the lateral intercellular spaces.

5.2.3   *Chloride reabsorption*

The intracellular electrical potential opposes Cl^- entry into the cells from the tubular lumen and in the early proximal tubule, Na^+ absorption is accompanied by HCO_3^- absorption (as CO_2 – as a result of H^+ secretion), so that although the tubular fluid osmolality is similar to that of plasma, the Cl^- concentration is increased. There are two main mechanisms whereby Cl^- is absorbed from the tubular lumen into the cells (Figure 5.1)

(1)   By antiport in exchange for the secretion of other anions. Some of this antiport may be Cl^- entry in exchange for HCO_3^- secretion. However, it is thought that an important anion exchanged for Cl^- is formate, $HCOO^-$. Thus Cl^- entry into the cells is accompanied by $HCOO^-$ entry to the tubular lumen. This $HCOO^-$ then combines with secreted H^+ (secreted by Na^+/H^+ antiport) to form $HCOOH$, which readily diffuses back into the cells, where it can dissociate into H^+ (for secretion in exchange for Na^+) and $HCOO^-$ (for secretion in exchange for Cl^-).
(2)   In the final two-thirds of the proximal tubule Cl^- handling is different in superficial nephrons than in juxtamedullary ones. In superficial nephrons

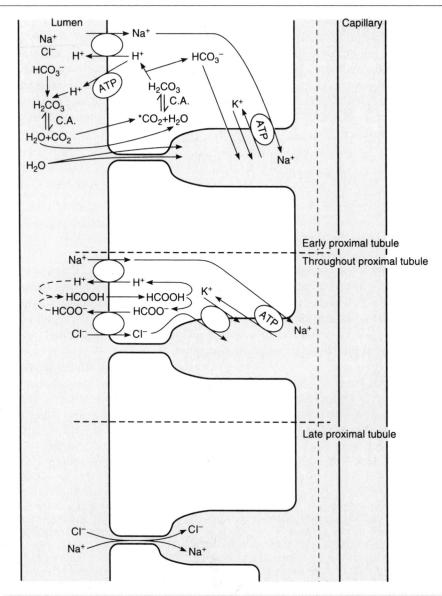

Figure 5.1 Sodium, chloride, bicarbonate and water movements across proximal tubule cells. The tubular lumen is on the left and the peritubular capillary is on the right. Primary active transport is shown as (ATP). Carrier-mediated symport and antiport processes are shown as ◯. The primary active transport process is the Na^+K^+-ATPase on the basal and basolateral cell membrane, which extrudes $3Na^+$ for $2K^+$ entering. The Na^+ concentration in the tubular lumen (and in the peritubular capillary) is 140 mM, whereas inside the tubule cell it is below 30 mM. Furthermore, the cell interior has a potential of -70 mV relative to the tubular lumen and the plasma. Hence Na^+ entry into the cell is favoured by both concentration and electrical gradients.

the late proximal tubule Cl^- permeability is greater than that for other anions (notably HCO_3^-). Hence Cl^- is reabsorbed down its concentration gradient (as a result of the high tubular Cl^- concentration mentioned above), with the consequence that Na^+ follows passively along its electrical gradient (in contrast, in the juxtamedullary nephrons, Cl^- and HCO_3^- permeabilities are not different). Up to 20% of NaCl reabsorption in the late proximal tubule of superficial nephrons is by this mechanism.

5.2.4   *Water reabsorption*

Some 60–70% of the filtered water is reabsorbed in the proximal tubules. The active extrusion of Na^+ from the tubular cells into the lateral intercellular spaces provides the driving force for this reabsorption of water, much of which passes through the cells via aquaporin-1 water channels present in the proximal tubule cell membranes (both apical and basolateral). There is some controversy about the precise mechanism of water reabsorption. Most evidence now indicates that reabsorption in the proximal tubule is not really isosmotic. In fact, solute reabsorption processes establish a small degree of hypotonicity (2–5 mosmol/kg H_2O) in the tubular lumen (relative to the plasma), which accounts for about 20% of water reabsorption. A further 40% of the water reabsorbed proximally is attributable to the different permeabilities of anions (primarily Cl^- and HCO_3^- described above; i.e. the fact that the tubular Cl^- concentration is increased by the end of the early part of the proximal tubule (see above) and that the later parts of the proximal tubule (in superficial nephrons) are more permeable to Cl^- than to other solutes (i.e. Cl^- is a relatively 'ineffective osmole', see p. 6), means that both Cl^- and water will be reabsorbed. The remaining 40% of the total proximal water reabsorption can

Figure 5.1   (Continued) H^+ secretion is mainly by antiport with Na^+, but there is also some secretion by H^+-ATPase. H^+ secretion is important for bicarbonate reabsorption (see Figure 5.6) and also for Cl^- reabsorption. Most Na^+ entry is by the Na^+/H^+ antiport. In the early part of the proximal tubule, more HCO_3^- is absorbed (Figure 5.6) than Cl^-, so the tubular Cl^- concentration increases, from an initial value of 115 mmol/l, to 130 mmol/l. Cl^- is then reabsorbed in the later part of the proximal tubule by the paracellular route and this absorbs some Na^+ by this same route. Cl^- is also absorbed by countertransport with organic anions. Formate ($HCOO^-$) is shown in the diagram, but oxalate may also be involved. Much of the active transport of Na^+ is directed into the lateral spaces, leading to the accumulation of anions and water in the spaces, as well as Na^+. The 'tight junction' between the tubule and the lateral intercellular space is not physiologically 'tight', and allows water and some ion movements. However, much of the water accompanying ion movements passes through the cells; it is able to do this at a high rate because of the presence of aquaporin-1 water channels in the cell membranes.

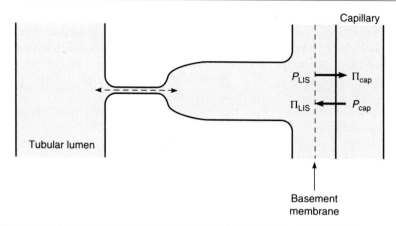

Figure 5.2 The forces (Starling forces) involved in the movement of fluid (water + solutes) from the lateral intercellular spaces into the peritubular capillaries. The transport of Na^+ from the cells into the lateral intercellular spaces, accompanied by negative ions (Cl^-, HCO_3^-), leads to the osmotic accumulation of water in the lateral spaces, generating a hydrostatic pressure in the lateral intercellular space (P_{LIS}) tending to force fluid into the capillary. The oncotic pressure in the capillary (Π_{cap}) also favours this capillary uptake. The forces opposing uptake are the capillary hydrostatic pressure (P_{cap}), and the lateral intercellular space oncotic pressure (Π_{LIS}).

be attributed to the existence of a hypertonic intermediate compartment (the lateral intercellular spaces – see Figure 5.2).

5.3 *Uptake of NaCl and water into peritubular capillaries*

The end result of the processes described so far is the entry of isotonic fluid (from the tubule) into the lateral intercellular spaces. From here, the NaCl/ $NaHCO_3$ solution can move in two possible directions: (1) into the capillaries, or (2) back into the tubular lumen.

A proportion of the Na^+ always leaks back, and hence the sodium reabsorption process can be called a 'pump-leak' system (it is also known as a gradient-time limited transport process). Alterations in the rate of proximal tubular sodium reabsorption can be brought about by changes in the rate of backflux into the tubule, as well as by changes in the rate of active sodium extrusion from the cells.

Changes in the rate of backflux occur as a result of changes in the rate of uptake from the lateral intercellular spaces into the capillaries – i.e. the faster the capillary uptake the lower the rate of backflux. What then determines the rate at which water and solutes are transferred from the lateral intercellular spaces into the capillaries?

The forces governing the movement of fluid across the walls of the peritubular capillaries are Starling forces – i.e. oncotic pressure and hydrostatic pressure gradients. So capillary uptake from the lateral intercellular spaces is determined as follows:

$$\text{Capillary uptake} \propto \text{Forces favouring uptake} - \text{Forces opposing uptake}$$
$$\propto (\Pi_{cap} + P_{LIS}) - (\Pi_{LIS} + P_{cap})$$

where Π_{cap} and Π_{LIS} are oncotic pressures in the peritubular capillary and the lateral intercellular space respectively (the latter is normally negligible) and P_{cap} and P_{LIS} are the hydrostatic pressures in the capillary and the lateral intercellular space (Figure 5.2).

5.4   *Relationship of proximal tubular reabsorption to glomerular filtration rate*

It is important at this stage to bear in mind that the peritubular capillaries are branches of the efferent arterioles, which in turn arise from the glomerular capillaries. Consequently, the Starling forces in the peritubular capillaries can be modified by the glomerular filtration process. The peritubular capillary oncotic pressure (Π_{cap}) depends on the plasma protein concentration. Since the proteins are concentrated in the *glomerular* capillaries by the filtration process, Π_{cap} in the peritubular capillaries is high and causes fluid reabsorption into the capillaries (see Figure 3.5). Peritubular capillary Π_{cap} will depend partly on the filtration fraction (GFR/RPF, see pp. 44 and 91).

The peritubular capillary hydrostatic pressure will be determined mainly by the venous pressure (see p. 167), but it is possible that changes in the degree to which arterial pressure is transmitted to the capillaries (via afferent and efferent arterioles) can also determine P_{cap}.

The dependence of Π_{cap} on the filtration fraction provides a mechanism whereby proximal tubular reabsorption could be adjusted automatically to compensate for changes in glomerular filtration: thus, if GFR increases, the forces available to reabsorb the increased volume of filtrate also increase, automatically. Thus there is **glomerulo-tubular balance**, whereby an essentially fixed percentage of the glomerular filtrate will be reabsorbed proximally, i.e. there is normally a relatively constant proximal tubular fractional reabsorption.

However, proximal tubular fractional reabsorption can be altered, e.g. by alterations in the effective circulating volume. Such alterations are considered later (p. 114).

5.5   *Proximal tubular reabsorption of other solutes*

The importance of sodium reabsorption in the proximal tubule derives not only from the necessity of conserving filtered sodium, but also from the effects of sodium reabsorption on the reabsorption of other solutes.

Sodium reabsorption leads to electrical, concentration and osmotic gradients for the passive reabsorption of such solutes as chloride, potassium and urea, and also for water absorption. It is also important for the reabsorption of glucose, amino acids, phosphate, calcium and bicarbonate, and for the secretion of H^+.

5.5.1 *Glucose*

In normal healthy people, almost all of the filtered glucose is reabsorbed and a negligible amount is excreted. Since the normal plasma glucose concentration is between 60 and 100 mg/100 ml (3.3–5.5 mmol/l) and the GFR is 125 ml/min, it is apparent that between

$$\frac{60 \times 125}{100} = 75\,\text{mg}\ (0.42\,\text{mmol})$$

and

$$\frac{100 \times 125}{100} = 125\,\text{mg}\ (0.7\,\text{mmol})$$

glucose is reabsorbed every minute.

Although most glucose reabsorption occurs in the proximal tubule, more distal parts of the nephron are also capable of reabsorbing glucose. Nevertheless, as the proximal tubule is undoubtedly the major site of glucose reabsorption, this chapter is the appropriate place to consider the glucose reabsorptive mechanism. Figure 5.3 shows the relationship between filtration, reabsorption and excretion of glucose, and the plasma glucose concentration.

The amount of glucose filtered is directly proportional to the plasma glucose concentration. In man, reabsorption of glucose is complete and none is excreted unless the plasma glucose concentration exceeds about 200 mg/ 100 ml (i.e., over 10 mmol/l). At this plasma glucose concentration, those nephrons with the lowest capacity for glucose reabsorption (relative to their filtration rate) reach their glucose reabsorptive rate limit, and glucose begins to be excreted. Further increases in plasma glucose concentration saturate the glucose transport process of an increasing proportion of nephrons until, when the plasma glucose concentration is about 400 mg/100 ml, no nephrons are able to absorb all of their filtered glucose load.

If all the nephrons had exactly the same glucose reabsorptive capacity (relative to their filtration rates), we might expect the curves for glucose reabsorption and excretion to be as shown in the dotted lines (Figure 5.3), with a sharp transition from zero excretion to a rate of excretion directly related to the plasma glucose concentration. The fact that the curves are 'splayed' indicates the existence of nephron heterogeneity, i.e. that all the nephrons are not identical.

The type of transport process typified by glucose reabsorption is known as T_m-limited transport. The term T_m means tubular maximum, and refers

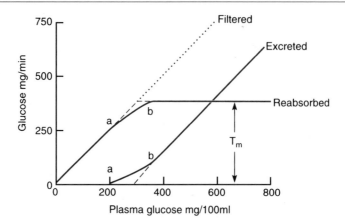

Figure 5.3 The relationship of glucose filtration, reabsorption and excretion to the plasma glucose concentration. The graph is for the normal glomerular filtration rate (125 ml/min). Thus if the plasma glucose concentration is 100 mg/100 ml (5.5 mmol/l), then 125 mg glucose will be filtered per minute. Until the plasma glucose concentration reaches about 11 mmol/l (200 mg/100 ml) all of the filtered glucose is reabsorbed and none is excreted. At this concentration (points **a** on the graph), those nephrons with the poorest ability to reabsorb glucose begin to allow the excretion of glucose. If the plasma glucose concentration rises further, even those nephrons with the greatest glucose reabsorptive capacity cannot reabsorb more (points **b** on the graph). Thus the 'splay' of the curves between points **a** and **b** indicates nephron inhomogeneity. If all the nephrons had identical glucose reabsorptive capacities relative to their filtration rate, the curves would be as shown by the dashed lines. The tubular transport maximum for glucose, T_m is about 380 mg/min, and is reached when plasma glucose concentration approaches 400 mg/100 ml (22 mmol/l).

to the maximum tubular reabsorptive capacity for a particular solute. From Figure 5.3 we can see that the maximum rate of glucose reabsorption is about 380 mg/min. This is the T_m for glucose.

Renal abnormalities of glucose excretion, leading to glycosuria, may occur either as a result of a reduced T_m for glucose, or because there is an abnormally wide range of nephron inhomogeneity (i.e. the 'splay' of the glucose excretion curve is increased). Renal glycosuria may occur transiently during pregnancy.

A more common abnormality of glucose excretion is that caused by a change in the plasma glucose concentration, so that the filtered load is altered: **diabetes mellitus** is caused by the relative or total absence of the pancreatic hormone, insulin, which regulates the blood glucose concentration. In the absence of insulin, the plasma glucose concentration increases, and can exceed 600 mg/100 ml plasma (33 mmol/l). The filtered load of glucose can therefore be far in excess of the reabsorptive capacity of the nephrons, so that glucose is excreted in the urine. This excretion of osmotically active solute causes an osmotic diuresis (p. 183), resulting in water loss from the body and hence dehydration and thirst.

5.5.2 *Relationship of glucose reabsorption to sodium reabsorption*

Proximal tubular glucose absorption is linked to sodium reabsorption, i.e. there is a symport (cotransport) process for the entry of glucose into the tubular cells from the lumen, whereby the passive entry of sodium into the cell down its electrochemical gradient permits the entry of glucose against its gradient. Thus glucose absorption is ultimately dependent on the maintenance of a gradient for the passive entry of sodium, which in turn depends on the extrusion of sodium from the peritubular side of the proximal tubule cells by the sodium pump (Figure 5.4).

In the convoluted segment of the proximal tubule (pars convoluta), the coupled sodium–glucose entry into the cells across the apical membrane occurs via a symporter termed SGLT-2 (Sodium GLucose Transporter 2), which transports one glucose molecule per Na^+ ion. In contrast, in the later, straight part of the

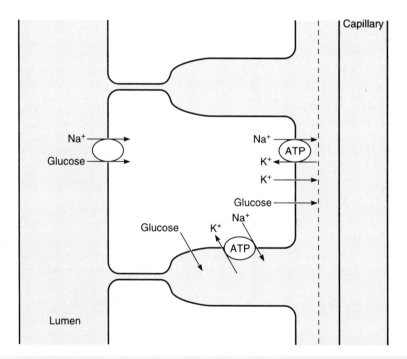

Figure 5.4 Symport (cotransport) of solutes with Na^+. The active extrusion of Na^+ on the basal (peritubular) side of the cell, by the Na^+ K^+-ATPase, generates a concentration (and electrical) gradient for Na^+ entry into the cell from the tubule, across the apical membrane. The cotransported substance enters along with Na^+. Symport with glucose is shown, but amino acids and phosphate are reabsorbed similarly. The entry of glucose (amino acids, phosphate) increases the intracellular concentration to provide a gradient for transport of the substance across the basal cell membrane. (Basal membrane glucose transport is carrier-mediated.)

proximal tubule (pars recta), the transporter (SGLT-1) transports one glucose molecule per two Na^+ ions (this transporter is also present in intestinal cells). The exit of glucose from the proximal tubular cells across the basolateral membrane is down its concentration gradient, and occurs by facilitated diffusion (see Chapter 4) via a transporter (uniporter) which does not require any other substance. This is GLUT-2 (GLUcose Transporter 2).

5.5.3 *Amino acids*

The plasma concentration of amino acids is 2.5–3.5 mmol/l. Amino acids in the plasma are in a dynamic equilibrium, since they enter the blood from the gut (as products of protein digestion) and are continually being used to restructure the body tissues.

Amino acids are small molecules and are readily filtered at the glomeruli. Negligible quantities of amino acids are excreted, however, because there are effective T_m-limited transport processes for amino acids in the proximal tubule. In fact, there are at least seven independent proximal transport processes for amino acid reabsorption.

These are for:

(1) basic amino acids and cystine;
(2) glutamic and aspartic acids;
(3) the neutral amino acids;
(4) imino acids;
(5) glycine;
(6) cystine and cysteine;
(7) Beta and gamma amino acids (e.g. taurine and gamma-aminobutyric acid – GABA).

The functional characteristics of these transport processes are very similar to that for glucose. Amino acid entry into the proximal tubule cells from the lumen is a cotransport process with sodium, the driving force being the sodium gradient (similar to the process for glucose shown in Figure 5.4).

5.5.4 *Phosphate*

Phosphate is an essential constituent of the body. Bones and teeth are salts of calcium and phosphate, and the skeleton accounts for about 80% of the body phosphate content. The other 20% is present mainly in intracellular fluid. The extracellular (plasma) phosphate concentration is 1 mmol/l (normally expressed as elemental phosphorus, see Chapter 12), and plasma phosphate is freely filtered at the glomerulus. In the nephron, tubular reabsorption and (possibly) secretion of phosphate occurs.

Normally, the urinary phosphate excretion is less than 20% of the amount filtered, but above a plasma phosphate concentration of about 1.2 mmol/l the

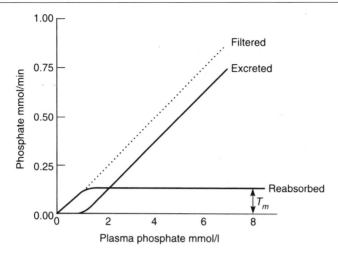

Figure 5.5 The relationship between the plasma phosphate concentration and the amount of phosphate filtered, reabsorbed and excreted. Although superficially similar to the curves for glucose (Figure 5.3), there is an important difference. Whereas the normal amount of glucose filtered is well below the T_m for glucose, the normal amount of phosphate filtered (which of course depends on the plasma concentration) is very close to the T_m for phosphate. Thus the reabsorption process in effect regulates the plasma phosphate concentration, since any increase in plasma phosphate concentration increases phosphate excretion.

increments in urinary excretion match the increments in filtration (Figure 5.5), suggesting that there is a T_m for phosphate. Micropuncture studies indicate that this T_m-limited reabsorption occurs in the proximal tubule, but the evidence remains somewhat controversial.

Phosphate reabsorption occurs only in the presence of sodium reabsorption, by an electroneutral cotransport (i.e. two Na^+ ions per phosphate ion) across the apical (brush-border) membrane. The basolateral membrane transport mechanism is still not clear. The rate of phosphate uptake is hormonally regulated, being under the control of PTH (parathyroid hormone) and vitamin D (Chapter 12).

5.5.5 *Urea*

The normal plasma urea concentration is 15–45 mg/100 ml (2.5–7.5 mmol/l). Whereas virtually 100% of filtered glucose is reabsorbed by the end of the nephron (mainly in the proximal tubule), only 40–50% of filtered urea is reabsorbed and 50–60% is excreted.

Urea is the end product of protein metabolism and, clinically, is measured as blood urea nitrogen (BUN) Because urea is a small molecule, it is reabsorbed in the proximal tubule as a consequence of sodium reabsorption. Thus, as sodium chloride and water are abstracted from the proximal tubule, the urea concentration in the tubular fluid tends to increase and so urea is reabsorbed passively by diffusing down its concentration gradient, out of the tubule.

The urea handling of the more distal parts of the nephron plays an important part in the process of concentrating the urine, and is considered later (Chapter 6).

5.5.6 *Bicarbonate*

The plasma HCO_3^- concentration is normally about 25 mmol/l. Bicarbonate is of great importance in the body because of its key role in acid–base balance, and the kidney contributes to acid–base balance largely by regulating the plasma bicarbonate concentration. How this regulation is accomplished is covered in detail in Chapter 10; at this stage we will consider the basic principles of renal bicarbonate handling.

About 90% of the filtered HCO_3^- is reabsorbed in the proximal tubule; the remainder is reabsorbed in the distal tubule and collecting ducts. In the proximal tubule the reabsorption occurs as a result of the secretion of H^+ from the cells into the tubular lumen. The reaction sequence involved is shown in Figure 5.6.

The bicarbonate reabsorption mechanism behaves as if there were a T_m for bicarbonate (Figure 5.7). However, the apparent T_m can be altered by the rate of H^+ secretion, which is itself loosely determined by the rate of Na^+ reabsorption. The exit of HCO_3^- from the proximal tubule cells into the peritubular interstitium is via an Na^+–HCO_3^- cotransporter (p. 129).

5.5.7 *Sulphate*

Sulphate is reabsorbed in the proximal tubule by an active T_m limited transport process which serves to maintain the plasma sulphate concentration at 1–1.5 mmol/l.

5.5.8 *Potassium*

(See Chapter 11)

5.5.9 *Albumin*

Albumin constitutes 60% of the total plasma protein, and is present in the plasma at a concentration of 45 g/l. The molecular weight of albumin is 68,000, and thus small amounts (0.05–0.1% of the amount delivered to the kidneys in

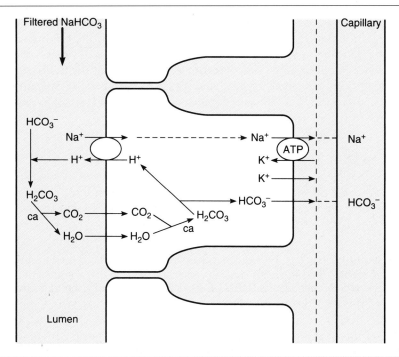

Figure 5.6 Bicarbonate reabsorption in the proximal tubule. Bicarbonate is filtered into the tubule (left of diagram) at the glomerulus. Na^+ enters the cells from the tubular lumen across the apical membrane along the concentration and electrical gradient generated by the Na^+K^+-ATPase on the basal membrane. In the early proximal tubule, most of the Na^+ entry into the cells from the tubular lumen is accompanied by H^+ secretion (see also Figure 5.1), which is mainly by countertransport with Na^+. The secreted H^+ combines with HCO_3^- to form carbonic acid (H_2CO_3), which forms CO_2 and H_2O. The formation of CO_2 and H_2O is catalysed by the enzyme carbonic anhydrase (ca), present in the brush border of the proximal tubule cells. Both CO_2 and H_2O can readily enter the tubule cells, where carbonic acid is again formed (catalysed by carbonic anhydrase) and dissociates into H^+ and HCO_3^-. The H^+ is secreted into the tubular lumen, and some of the HCO_3^- enters the blood together with reabsorbed Na^+ (right of diagram). However, some of the HCO_3^- generated in the cells exchanges with luminal Cl^-. The end result is that filtered $NaHCO_3$ and $NaCl$ is reabsorbed and H^+ recycles between the tubular lumen and the cells.

the plasma) are filtered, as described in Chapter 3. This means that the daily filtered load of albumin is about 8 g. Since the plasma volume is 3 l. The total amount of albumin in this is 45×3 g, i.e. 135 g, and the daily filtered load, 8 g, is 6% of the total albumin in the body. A loss of albumin on this scale would have to be countered by an increase in dietary protein intake. Such a loss does not happen, because albumin is reabsorbed in the proximal tubule.

The albumin reabsorption occurs by a mechanism called *adsorptive endocytosis* or *receptor mediated endocytosis*. The albumin in the tubule binds to specific

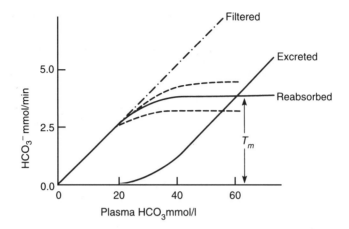

Figure 5.7   The relationship between the plasma bicarbonate concentration and the amount of bicarbonate filtered, reabsorbed and excreted (at a GFR of 125 ml/min). The T_m for bicarbonate is variable (as shown by dashed lines), so that the amount excreted at a given plasma concentration also varies (although for clarity the variability in the amount excreted is not shown).

sites on the apical membrane of the proximal tubule cells, and these sites are then endocytosed into the cells, and delivered to lysosomes where the albumin is cleaved to amino acids, which leave the cell across the basolateral membrane. The binding sites become reincorporated in the apical membrane.

5.6   *Secretory processes in the proximal tubule*

Tubular secretory mechanisms are very much like tubular reabsorptive mechanisms – the important difference being the direction of transport. Secretion is movement into the tubule, whereas reabsorption is movement from the tubule (and thence to the peritubular fluid). Like reabsorptive processes, secretion can be either active or passive – and secretory processes may be gradient-time limited (e.g. the proximal tubule hydrogen secretion), or T_m limited.

There are three proximal tubular secretory mechanisms which have a definite T_m limit. These are for:

(1)   a group of organic acids: these include penicillin, chlorothiazide, hippurate, *p*-aminohippurate (PAH) and possibly uric acid (Figure 5.9);
(2)   a group of strong organic bases, which includes histamine, choline, thiamine, guanidine and probably creatinine and tetraethylammonium;
(3)   EDTA (ethylene diamine tetraacetic acid).

Although most of these substances do not occur endogenously, endogenous organic acids are also transported.

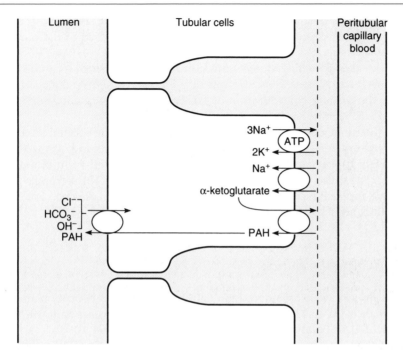

Figure 5.8 The mechanism of organic acid secretion in the proximal tubule (using PAH as the example of an organic acid). PAH entry into the cells across the basal membrane is by counter-transport (antiport) with di- or tri-carboxylic acids, such as α-ketoglutarate (as shown). PAH exit from the cells into the tubular lumen is by antiport with an anion (HCO_3^-, Cl^- or OH^-).

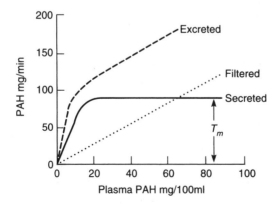

Figure 5.9 The filtration, secretion and excretion of PAH, as a function of the plasma PAH concentration. For further explanation see Figure 7.2.

There is some anatomical separation of the different secretory processes. Organic acid secretion (and uric acid secretion) takes place in the pars recta, but organic base secretion occurs in the pars convoluta.

PAH, for the physiologist, is a very useful substance, since it can be used to determine renal plasma flow (Chapter 7). Figure 5.8 shows the relationship between the plasma concentration of PAH and the rates of filtration, tubular secretion and urinary excretion.

Over the range of plasma concentrations from 0 to 8 mg/100 ml, PAH secretion can completely remove PAH from the tubular capillaries (thus the only PAH which appears in the renal venous blood is derived from blood which did not go past the proximal tubules). The T_m for the PAH secretory process occurs at a plasma concentration of 10–20 mg/100 ml plasma. For further explanation, see Chapter 7, Figure 7.2.

5.6.1 *Hydrogen secretion*

This has already been considered in relation to sodium reabsorption and bicarbonate reabsorption. The relationship of hydrogen ion secretion to the regulation of acid–base balance is considered in detail in Chapter 10.

Further reading

Alpern, R. J., Moe, O. W. and Preisig, P. A. (1995) Chronic regulation of the proximal tubular Na/H antiporter. *Kidney Int.* **48**, 1386–1396

Aukland, K., Bogusky, R. T. and Renkin, E. M. (1994) Renal cortical interstitium and fluid absorption by peritubular capillaries. *Am. J. Physiol.* **266**, F175–F184

Berry, C. A. and Verkman, A. S. (1988) Osmotic gradient dependence of osmotic water permeability in rabbit proximal convoluted tubule. *J. Membr. Biol.* **105**, 33–43

Dantzler, W. H., Wright, S. H. and Lote, C. J. (1997) Organic solute transport. In Jamison, R. L. and Wilkinson, R. (eds), *Nephrology*, Chapman and Hall, London, pp. 61–70

Kaplan, M. R., Gamba, G. and Hebert, S. C. (1996) Molecular mechanisms of NaCl transport. *Annu. Rev. Physiol.* **58**, 649–668

Wakabayashi, S., Shigekawa, M. and Pouyssegur, J. (1997) Molecular physiology of vertebrate Na^+/H^+ exchangers. *Physiol. Reviews* **77**, 51–74

Wang, T., Egbert, A. L., Abbiati, T., Aronson, P. S. and Giebisch, G. (1996) Mechanisms of stimulation of proximal tubule chloride transport by formate and oxalate. *Am. J. Physiol.* **271**, F446–F450

Zelikovic, I. and Chesney, R. W. (1989) Sodium-coupled amino acid transport in renal tubule. *Kidney Int.* **36**, 351–359

Problem

5.1 A subject has a plasma glucose concentration of 4.2 mM (76 mg/100 ml). His glomerular filtration rate is found to be 110 ml/min.

(a) What is the filtered load of glucose?
(b) Would you expect glucose to be present in the urine of this subject?

The loop of Henle, distal tubule and collecting duct
6

6.1 *The loop of Henle*

The fluid entering the loop of Henle is isotonic to plasma, but, after traversing the loop, fluid entering the distal tubule is hypotonic to plasma, i.e. the tubular fluid has been diluted during its passage around the loop of Henle. Only mammals and birds are able to produce concentrated urine (i.e. urine hypertonic to plasma), and only mammals and birds have loops of Henle.

This paradoxical situation – that a nephron segment which dilutes the tubular fluid determines the maximum possible urinary osmolality – was a puzzle to renal physiologists for many years. The solution to the paradox came with the realization that the loops of Henle are **countercurrent multipliers**, the function of which is not to concentrate the tubular fluid within them, but to manufacture a hypertonic interstitial fluid in the renal medulla. Urine is then concentrated by the osmotic abstraction of water from the collecting ducts as they pass through the medulla.

6.1.1 *Countercurrent multiplication mechanism*

The theory for the mechanism of countercurrent multiplication in the loop of Henle was propounded by Wirz, Hargitay and Kuhn in 1951. Essentially, the theory proposed that if the loop of Henle was able to produce a small osmotic pressure difference between the ascending and descending limbs of the loop (i.e. a small **transverse gradient**), then this small difference would be multiplied into a large **longitudinal gradient** by the countercurrent arrangement (i.e. flow in opposite directions) in the two limbs of the loop.

This theory is now known to be essentially correct. The ascending limb of the loop actively extrudes Na^+ into the medullary interstitium, but is impermeable to water, so that water is unable osmotically to follow the Na^+ and accompanying ions (Figure 6.1). Consequently, the osmolality of the medullary interstitium is increased and the osmolality of the fluid in the ascending limb is decreased.

The primary active (i.e. energy-consuming) transport process in the (thick) ascending limb of the loop of Henle is the Na^+K^+-ATPase on the basal cell membrane. The entry of solutes into the cell across the apical membrane involves cotransport of sodium, chloride and potassium, with the stoichiometry of one Na^+, two Cl^- and one K^+, so the process is electrically neutral. This transport can be inhibited by loop diuretics, such as furosemide, bumetanide and piretanide (see Chapter 15). Much of the K^+ leaks back into the tubular lumen, so that it is predominantly NaCl which accumulates in the medullary interstitium. The ascending limb is impermeable to water, so that water is unable osmotically to follow the ions. Consequently, the osmolality of the medullary interstitium is increased and the osmolality of the fluid in the ascending limb is decreased. The tubular transport processes are shown in Figure 6.2.

To examine countercurrent multiplication in detail, let us imagine that we can 'turn off' the active processes which occur in the loop of Henle, then fill the loop with isotonic fluid from the proximal tubule. This fluid is actually of osmolality 285 mosmol/kg H_2O, but is shown as 290 in Figure 6.3 to simplify the arithmetic!

The cells in the wall of the ascending limb are able to sustain an osmotic pressure difference between the luminal side and interstitial side (i.e. a transverse gradient) of about 200 mosmol/kg H_2O. So, if the fluid in the ascending limb in our hypothetical situation (before we turn on the pump) has an osmolality of 290 mosmol/kg H_2O, the pump can lower the luminal osmolality to 190 mosmol/kg H_2O and raise the interstitial osmolality to 390 mosmol/kg H_2O. This sequence of events is shown in Figure 6.3.

The descending limb is permeable to water and, to a lesser extent, is also permeable to NaCl. The fluid within the descending limb will therefore come to osmotic equilibrium with the interstitium, partly by water moving out of the descending limb, and partly by NaCl moving in. In effect, then, we can consider the transport of NaCl out of the ascending limb as being directed into the descending limb.

With the above, we are now in possession of all the facts necessary to understand how the counter current multiplication system works and, in Figure 6.3, the process is artificially divided into a series of separate stages. In stage 1, the loop is filled up with isotonic fluid from the proximal tubule. In stage 2, we operate the NaCl pump in the ascending limb, which produces a gradient of 200 mosmol/kg H_2O between the ascending and the descending limb. If we now (stage 3) move more isotonic fluid in from the proximal tubule and move some of the hypertonic fluid around the bend from the descending

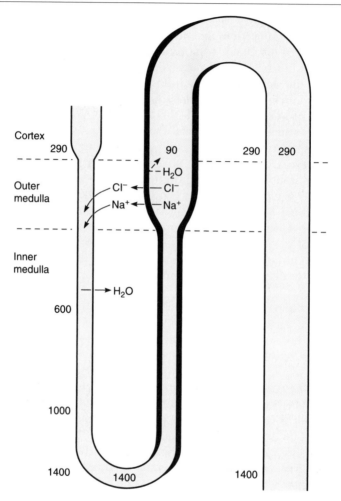

Figure 6.1 The fundamental transport processes in the loop of Henle, responsible for the establishment and maintenance of the hypertonicity of the medullary interstitium. Na^+ is actively transported from the (thick) ascending limb of the loop of Henle, and this leads to the reabsorption of Cl^- (for details of the cellular transport mechanism, see Figure 6.2). This segment of the nephron is impermeable to H_2O and the osmolality of the fluid in the ascending limb therefore falls, and that of the medullary interstitium increases. The descending limb of Henle's loop is permeable to both ions and water, and so comes into osmotic equilibrium with the medullary interstitium. This occurs partly by the movement of NaCl from interstitium to descending limb, and partly by movement of water from descending limb to interstitium. The transition from water permeability to water impermeability of the tubule is thought to occur just before the tip of the loop of Henle. The figures in the diagram show osmolality, mosmol/kg H_2O.

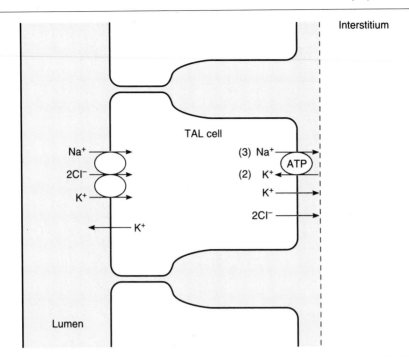

Figure 6.2 Ion transport in cells of thick ascending limb (TAL) of the loop of Henle. These cells are almost impermeable to water. The entry of Na^+, K^+ and Cl^- into the cells from the tubular lumen is by a symport (cotransport) process, driven by the Na^+ gradient. The stoichiometry is 1 Na^+, 2 Cl^-, 1 K^+. On the basal (peritubular) side of the cell, Na^+ is pumped out by the Na^+K^+-ATPase, and Cl^- follows passively. The cells are able to sustain an osmolality difference of 200 mosmol/kg H_2O between the lumen and the interstitium. Note, in the diagram as shown, that 2 Cl^- are translocated from lumen to interstitium per 1 Na^+, with little net movement of K^+. The electrical imbalance which this could engender is prevented by backdiffusion of Cl^-, and the absorption of positive ions (including Na^+, Mg^{2+}, Ca^{2+}), between the cells.

to the ascending limb, we can establish a new equilibrium (stage 4) by opera-tion of the NaCl pump. As this sequence continues (stages 5 to 8), the osmolal-ity at the tip of the loop (and in the interstitium) progressively increases. In man, the osmolality at the papilla tip can reach 1400 mosmol/kg H_2O, i.e. about five times the plasma osmolality. So, with the counter current multi-plier, the small transverse gradient (200 mosmol/kg H_2O) can be converted into a much larger longitudinal gradient.

So far, a number of details have been omitted from this scheme. The struc-ture of the ascending limb of the loop of Henle is not uniform, there being a thin segment and a thick segment (Figure 6.1). Both segments are imperme-able to water, but it may be that only the thick segment actively extrudes NaCl. The thin ascending limb has virtually no Na^+K^+-ATPase activity.

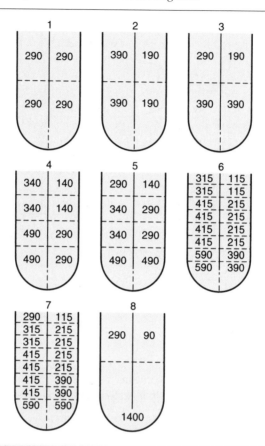

Figure 6.3 Sequence of events in the establishment of medullary hypertonicity by countercurrent multiplication. It is assumed that the ascending limb of the loop of Henle can produce an osmotic difference of 200 mosmol/kg H_2O between the tubular lumen and the medullary interstitium. The descending limb attains the same osmolality as the medullary interstitium and we can therefore regard the transport out of the ascending limb as being directed into the descending limb. In each diagram, the descending limb is on the left and the ascending limb is on the right.

(1) Assume that we can start by having the loop of Henle transport processes 'switched off', so that we can fill up the loop with fluid of osmolality 290 mosmol/kg H_2O from the proximal tubule.

(2) If we now operate the NaCl pump in the ascending limb, the pump produces an osmotic difference of 200 mosmol/kg H_2O at each transverse level in the medulla, i.e. the osmolality of the fluid in the ascending limb falls by 100 mosmol/kg H_2O and the osmolality of the fluid in the descending limb rises by 100 mosmol/kg H_2O.

(3) We now move the contents of the loop round, ejecting fluid into the distal tubule and introducing more fluid of osmolality 290 mosmol/kg H_2O from the proximal tubule.

(4) In this diagram, our previous four segments of the loop have been subdivided to make eight segments. The pump has again been operated to maintain a difference of 200 mosmol/kg H_2O between the ascending and descending limbs at any level.

However, this does not affect the basic principle of the countercurrent arrangement, which still depends on the removal of NaCl from the ascending limb, unaccompanied by water.

As a result of the countercurrent multiplication mechanism, the fluid in the tubules leaving the medulla and entering the cortex is hypotonic to plasma, with an osmolality of as low as about 90 mosmol/kg H_2O. Thus the ascending limb of the loop of Henle (and its continuation in the cortex as the early part of the anatomically defined 'distal tubule') can be called the 'diluting segment' of the nephron. (However, the nephrons with short loops of Henle, and hence a shorter diluting segment, are unlikely to lower the osmolality of the ascending limb fluid to 100 mosmol/kg H_2O.)

The fact that the 'distal tubule' is not a distinct section of the nephron physiologically is responsible for the considerable confusion concerning the functions of this segment. The problem arises from the fact that different species, different strains of animals, and even different individuals, vary in the point at which ascending-limb-of-Henle-type cells are replaced by collecting-tubule-type cells. The walls of the former cell type have only a low (and essentially constant) permeability to water, whereas the walls of the collecting-tubule-type cells have a variable water permeability, regulated by the hormone ADH (antidiuretic hormone, vasopressin).

The potential difference across the distal tubular region varies with distance along the tubule. In the early part, the lumen is positive (as in the ascending limb of Henle), but in the later parts the luminal potential (relative to plasma) is negative and may reach $-45\,mV$. This negative potential is caused by active sodium reabsorption.

Figure 6.3 (Continued)
(5) Ejection of more fluid into the distal tubule and introduction into the descending limb of fluid with osmolality 290 mosmol/kg H_2O from the proximal tubule. The fluid at the tip of the loop is moved round so that fluid with osmolality 490 mosmol/kg H_2O is present at the bottom of both descending and ascending limbs.
(6) Each segment has again been subdivided, and the pump operated to produce a 200 mosmol/kg H_2O difference between ascending and descending limbs at any level.
(7) Further flow around the loop so that fluid of osmolality 590 mosmol/kg H_2O is present at the bottom of both limbs.
(8) Gradient fully established. Fluid entering the descending limb from the proximal tubule has an osmolality of 290 mosmol/kg H_2O. Fluid leaving the ascending limb into the distal tubule has an osmolality of about 90 mosmol/kg H_2O. The osmolaity at the tip of the loop is 1400 mosmol/kg, but the difference in osmolalities between ascending and descending limbs at any transverse level is only 200 mosmol/kg H_2O. The interstitial osmolality is the same as that of the descending limb.

6.2   *Collecting tubules*

The collecting tubules have cortical and medullary sections, and the two sections have somewhat different properties.

The cortical and medullary sections are both relatively impermeable to water, urea and NaCl, but the **water permeability** is increased by ADH (for mechanism, see Chapter 8). Thus, ADH leads to urine concentration by permitting the osmotic abstraction of water into the interstitium, so that the urine in the collecting tubule can theoretically achieve the same osmolality (up to 1400 mosmol/kg H_2O) as the medullary interstitium, although it is usually rather less than this. Water reabsorbed from the medullary collecting tubules will tend to dilute the medullary interstitium, so that, for the concentration process to remain effective, the fluid delivery to the medullary collecting tubule (i.e. the fluid available for reabsorption in the medulla) must be small in volume. This is brought about (under the influence of ADH) by water reabsorption in the cortical collecting tubules. It was mentioned above that the osmolality of the fluid leaving the distal tubule can be as low as 90 mosmol/kg H_2O. (The volume of tubular fluid at this stage is normally not more than 20% of the glomerular filtrate, the rest having already been absorbed in the proximal tubule and loop of Henle.) In the presence of ADH, water reabsorption in the cortical collecting tubule will account for up to 66% of the fluid delivered to it (i.e. it will be concentrated from 90 to 290 mosmol/kg H_2O by osmotic abstraction of water) leaving less than 5% of the glomerular filtrate to continue into the medullary collecting tubules.

ADH increases the **urea permeability** of the medullary collecting tubules, but has no effect on the urea permeability of the cortical collecting tubules. This impermeability of the cortical part of the collecting tubule is one of the factors that make urea so important in the urine concentration mechanism.

6.3   *Importance of urea in countercurrent multiplication*

Although we have so far considered the countercurrent multiplication process only in terms of NaCl transport into the interstitium, a considerable fraction of the interstitial osmotic pressure is attributable to urea. The normal plasma concentration of urea is 15–45 mg/100 ml (i.e. 2.5–7.5 mmol/l). Urea is freely filterable at the glomerulus, and approximately 50% of the filtered load is reabsorbed (passively) in the proximal tubule. As the tubular fluid passes down the descending limb of the loop of Henle, the urea concentration increases, as a result of the diffusion of urea from the interstitium, down a concentration gradient, into the tubule (why there is such a high urea concentration in the medullary interstitium will soon be apparent).

When the tubular fluid reaches the 'distal tubule' and cortical collecting tubule, the urea concentration rises still more, as a result of water reabsorption, since these segments are almost impermeable to urea. When the medullary collecting tubule is reached, the high urea concentration in the tubule causes

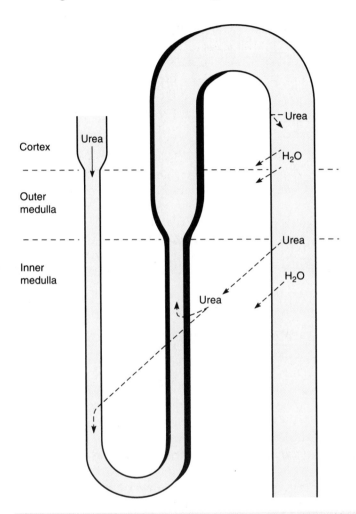

Figure 6.4 Trapping of urea in the medullary interstitium. Urea is delivered to the loop of Henle from the proximal tubule, passes round the loop of Henle to enter the distal tubule and thence the collecting ducts. In the presence of ADH, water is reabsorbed from the cortical collecting ducts, but this segment is impermeable to urea and therefore the luminal urea concentration increases. In the medullary collecting ducts, in the presence of ADH, both urea and water can leave the tubule and urea diffuses down its concentration gradient into the medullary interstitium. The high interstitial urea concentration thus achieved leads to the diffusion of some urea into the loop of Henle, to return to the collecting duct.

the diffusion of urea out of the tubule into the interstitium (the medullary collecting tubule is permeable to urea in the presence of ADH), thereby raising the interstitial urea concentration and so causing diffusion of urea into the descending limb of the loop of Henle. Thus urea recycles in the medulla, as shown in Figure 6.4. The effect of ADH on the urea permeability of the medullary collecting tubule is due to the fact that ADH causes the activation, in both the apical and basolateral membranes of the collecting tubule cells, of a **urea uniporter** – i.e. there is facilitated diffusion of urea out of the nephron at this site.

Since the papillary osmolality is due in part (up to 50%) to urea, the papillary osmolality is increased by increasing ADH levels, because in the presence of ADH the medullary collecting tubule is permeable to urea so that urea diffuses into the interstitium. Also, the maximum attainable urinary osmolality is greater when urea excretion is high than when urea excretion is low (urea is a product of protein metabolism, so a high protein diet increases urea excretion), because a proportion of this extra urea enters the medullary interstitium.

6.4 *Further requirements of the countercurrent multiplication mechanism: the vasa recta*

The vasa recta are capillaries, mostly derived from the efferent arterioles of juxtamedullary nephrons, which have a 'hairpin' arrangement and dip far down into the renal medulla.

In parts of the body other than the renal medulla, the circulation of blood through capillaries (which are freely permeable to water and small molecules) ensures the uniformity of the composition of the interstitial fluid. If the medullary capillaries served a similar purpose, the osmotic gradient built up by the loops of Henle would be dissipated. This does not occur, because the hairpin arrangement of the vasa recta enables them to function as **countercurrent exchangers** (distinguished from countercurrent multipliers by the fact that no energy is necessary). Thus, the function of the vasa recta is to provide nutrients to, and remove waste products from, the renal medulla, without washing away the solutes responsible for medullary hypertonicity.

The vasa recta, like capillaries elsewhere, are permeable to water and solutes. So, as plasma in the descending vasa recta passes down into the medulla, water will be osmotically abstracted from the vasa recta into the interstitium and solutes (NaCl and urea) will enter (Figure 6.5). More and more water will come out as the blood passes further into the medulla, until, at the tip of the loop, the plasma has almost the same osmolality as the surrounding interstitium and the blood is very viscous (with a high plasma protein concentration as a result of water loss to the interstitium). In the ascending vasa recta, the plasma regains the water and loses most of the solutes.

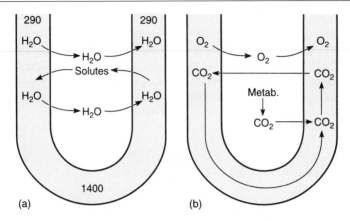

Figure 6.5 Counter current exchange in the vasa recta. (a) As the descending vasa recta enter the increasingly hypertonic medullary interstitium, water is osmotically abstracted from the blood vessel, so that the osmolality (in mosmol/kg water) of the blood (and its viscosity) are increased. In the ascending limb, water re-enters the blood vessel. The system ensures a low flow rate through the deep parts of the vasa recta, and minimizes the washout of medullary solutes. (b) O_2 and CO_2 also undergo counter current exchange in the vasa recta, so that the vasa recta are rather inefficient suppliers of O_2 and removers of CO_2 for cells deep in the medulla.

Some solutes are, however, washed out from the medulla, but if this were not the case, there might be no limit to how hypertonic the interstitium could become (i.e. theoretically counter current multiplication could go on *ad infinitum*). Other limiting factors for the degree of hypertonicity attained in the interstitium are the volumes of water reabsorbed from the collecting ducts and descending loops of Henle (most of this water enters the vasa recta and is removed from the medulla), and diffusion of solutes longitudinally within the medullary interstitium, towards the cortex.

The loop arrangement of the vasa recta renders them very inefficient at delivering oxygen to, and removing carbon dioxide from, the renal medulla. Some reports have suggested that the formation of ATP (to power the Na^+K^+-ATPase) in the loop of Henle is dependent on glycolysis rather than aerobic metabolism.

6.5 *Long and short loops of Henle*

It has been mentioned that, in man, only 15% of the nephrons (the juxtamedullary nephrons) have long loops of Henle which pass deeply into the medulla (Chapter 2). The remaining 85% of nephrons (cortical nephrons) have short loops of Henle which barely reach the medulla.

The nephrons with short loops of Henle do not make a significant contribution to the manufacture of medullary hypertonicity. However, the collecting tubules of **all** the nephrons (both cortical and juxtamedullary), pass through the medulla; thus, the long-looped nephrons, which are 15% of the total, produce a medullary gradient which leads to the concentration of urine from all the nephrons.

6.6   *Regulation of urine concentration*

The urine osmolality can range from about 60 mosmol/kg H_2O up to 1400 mosmol/kg H_2O, and the volume per 24 hours can be as little as 400 ml or as much as 23 l. How are such dramatic changes brought about?

The main determinant of whether the urine will be copious and dilute, or small in volume and concentrated, is the level of circulating ADH (antidiuretic hormone, vasopressin). The ways in which the level of ADH is regulated are considered later (Chapters 8 and 9). At this stage, we will simply look at the effects of altering the level of circulating ADH. This section will also serve to summarize the information we have covered so far concerning urine production.

6.6.1   *Maximal ADH: the production of concentrated urine (Figure 6.6)*

The reabsorption of water in the nephron occurs at several sites. The glomerular filtration rate is 180 l/day and approximately 70% of this is reabsorbed in the proximal tubule, so that about 53 l/day of **isotonic** fluid is delivered to the loops of Henle.

A further 15% (approximately 30 l/day) of the glomerular filtrate is reabsorbed by the loop of Henle (descending limb) leaving 23 l/day to enter the 'distal tubule'. The early 'distal tubule' reabsorbs very little water leaving 23 l to enter the late distal tubule and collecting tubules. It is this 23 l which can be either mainly excreted or mainly reabsorbed depending on the level of circulating ADH. In the presence of ADH, the collecting tubule wall is permeable to water, so water reabsorption occurs. Cortical absorption accounts for up to 66% of the water entering the collecting tubules (in the presence of ADH; see Section 6.2), so the delivery of fluid to the medullary collecting tubule is low (8 l/day, approximately) and, as this fluid passes down the tubule, the hypertonic medullary interstitium leads to the osmotic abstraction of water from the tubule. Urea also leaves the tubule in this region. The urine volume in these circumstances can be as little as 400 ml/day.

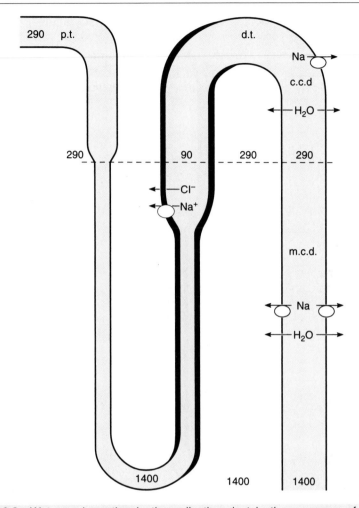

Figure 6.6 Water reabsorption in the collecting duct in the presence of maximal plasma levels of ADH. The figures show tubular fluid osmolality in mosmol/kg H_2O. Abbreviations: p.t., proximal tubule; l.h., loop of Henle; d.t., distal tubule; c.c.d., cortical collecting duct; m.c.d., medullary collecting duct. In the proximal tubule, water and solute reabsorption occur, but the tubular fluid remains isotonic to plasma (osmolality 290 mosmol/kg H_2O). In the loop of Henle, counter current multiplication produces an osmolality at the tip of the loop of 1400 mosmol/kg H_2O. Fluid entering the distal tubule is hypotonic. Some water is absorbed in the distal tubule, but sodium chloride reabsorption also occurs and the tubular fluid remains hypotonic (90 mosmol/kg H_2O). In the cortical collecting duct, water absorption occurs in the presence of ADH and the tubular fluid becomes isotonic to plasma (290 mosmol/kg H_2O); the volume of fluid delivered to the medullary collecting duct is small and water absorption along the osmotic gradient into the medullary interstitium raises the tubular fluid osmolality to close to 1400 mosmol/kg H_2O. Note that some sodium absorption occurs in the collecting ducts.

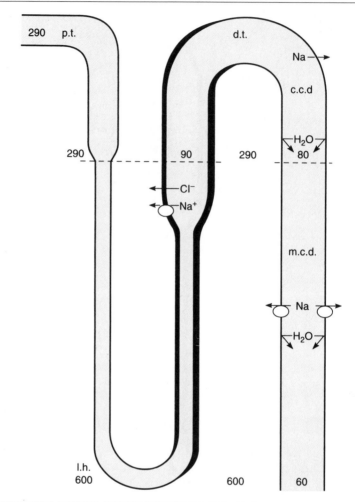

Figure 6.7 Water reabsorption in the collecting duct in the absence of ADH. For abbreviations, see Figure 6.6. Events in the proximal tubule are identical to those in Figure 6.6. In the loop of Henle, the tubular fluid becomes hypertonic, but the gradient is lower than in the presence of ADH (see text). Distal tubular events are as in Figure 6.6. In the cortical collecting duct, in the absence of ADH, water reabsorption does not occur. Sodium reabsorption therefore tends to lower the tubular fluid osmolality still further. This process continues in the medullary collecting duct and a large volume of very dilute urine is produced.

6.6.2   *No circulating ADH: the production of dilute urine (Figure 6.7)*

In the absence of circulating ADH, water reabsorption in the proximal and distal tubules occurs as above – i.e. ADH does not affect water absorption at these

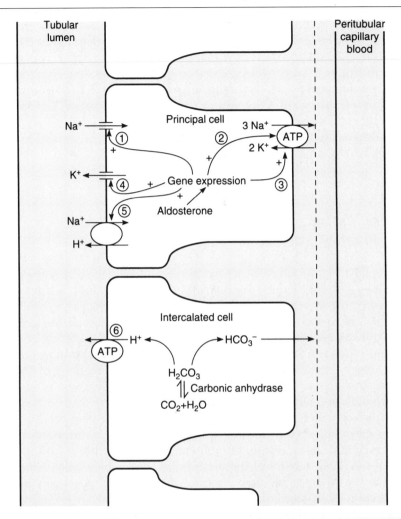

Figure 6.8 The actions of aldosterone in the distal nephron. By its action on gene expression in principal cells, aldosterone increases ①. functional Na^+ channels in the apical membrane; ②. the rate of ATP turnover; ③. the synthesis of Na^+ K^+-ATPase molecules for the basolateral membrane; ④. the number of K^+ channels in the apical membrane; this enhances K^+ secretion; ⑤. Na^+H^+ exchange. In intercalated cells, aldosterone increases ⑥. H^+ ATPase in the apical membrane. Actions 1–3 and 5 enhance sodium reabsorption, which also (coupled with action ④) increases potassium secretion, and (coupled with action ⑥) increases hydrogen ion secretion.

sites. However, the impermeability of the collecting tubules in the absence of ADH means that a large volume of water enters the medullary collecting tubule and is excreted. In addition, the impermeability of medullary collecting tubules to urea prevents the attainment of maximal medullary interstitial tonicity, and this in turn may slightly reduce water reabsorption from the descending limb of the loop of Henle. So ADH not only determines the urine volume, but also influences the medullary tonicity.

Normally, there is some ADH present in the circulation and the urine volume is approximately 1.5 l/day, with an osmolality of 300–800 mosmol/kg H_2O, but in the absence of ADH urine volume is approximately 23 l/day, with an osmolality as low as 60 mosmol/kg H_2O.

It should be noted that adrenal steroids (cortisol) must be present for ADH to have its maximum effect on water permeability.

6.7   *Other hormones affecting distal nephron transport processes*

Apart from ADH, the other major hormone affecting the distal nephron is aldosterone, a steroid synthesized in the zona glomerulosa of the adrenal gland from cholesterol. The control of aldosterone release is covered in Chapter 9. The main site of action of aldosterone is the cortical collecting tubule. Aldosterone diffuses into the principal cells of the cortical collecting tubule and binds to a specific cytosolic receptor to create a hormone–receptor complex which enters the cell nucleus; here it enhances messenger RNA and ribosomal RNA transcription, which in turn induces the synthesis of protein(s) that are thought to mediate the effects of the hormone.

The primary effect of aldosterone is to increase the permeability of the collecting tubule luminal membrane to Na^+ (and K^+). This increases the rate of Na^+ entry into the cells from the tubular lumen, and this in turn increases the Na^+K^+-ATPase activity of the peritubular (basal) membrane (since Na^+ delivery to the ATPase is the rate-limiting step). Aldosterone also stimulates the production of ATP, and of Na^+K^+-ATPase (Figure 6.8).

Aldosterone also increases the reabsorption of Cl^-, and the secretion of H^+ and K^+. The effect on Cl^- is secondary to that on Na^+, but there is evidence that the effects on H^+ and K^+ are independent of those on Na^+.

A number of other hormones or humoral agents influence ion transport in the distal nephron. Such agents include prostaglandins (p. 116), atrial natriuretic peptides (p. 117) and natriuretic hormone (p. 120).

Further reading

Berl, T. (1998) Water channels in health and disease. *Kidney Int.* **53**, 1417–1418
Jamison, R. L. (1987) Short and long loop nephrons. *Kidney Int.* **31**, 597–605

Reeves, W. B. and Andreoli, T. E. (1992) Sodium transport in the loop of Henle. In D. W. Seldin and C. Giebisch (eds), *The kidney, physiology and pathophysiology*, 2nd edition, Raven Press, New York, pp. 1975–2001

Wang, X., Thomas, S. R. and Wexler, A. S. (1998) Outer medullary anatomy and the urine concentrating mechanism. *Am. J. Physiol.* **274**, F413–F424

Zeidel, M. L. (1996) Low permeabilities of apical membranes of barrier epithelia: what makes watertight membranes watertight? *Am. J. Physiol.* **271**, F243–F245

Renal blood flow and glomerular filtration rate

7

Since the functioning of the kidneys depends on filtration of the plasma, the blood flow to the kidneys is of obvious importance. How the blood flow and glomerular filtration rate are regulated is the subject of the second part of this chapter. First, the measurement of renal blood flow (RBF) and glomerular filtration rate (GFR) will be considered.

7.1 Measurement of renal blood flow and glomerular filtration rate

In order to discuss the ways in which renal blood flow and glomerular filtration rate can be measured, the concept of **clearance** must be introduced. The clearance of any substance excreted by the kidney is the **volume of plasma** which is cleared of the substance in unit time. The units of clearance are volume/time, usually ml/min.

Consider the clearance of a substance, x. Clearance is given by the formula:

$$C_x = \frac{U_x V}{P_x}$$

where C_x is the clearance of x, U_x is the urine concentration of x, P_x is the plasma concentration of x and V is the urine flow (ml/min). If we express the formula in terms of the units of measurement

$$C_x = \frac{U_x \, (\text{mg/ml}) \times V \, (\text{ml/min})}{P_x \, (\text{mg/ml})}$$

it should be obvious that the units of clearance are ml/min.

In fact, the clearance represents a theoretical volume of plasma which is completely cleared of the substance, x, in 1 min, because in reality no aliquot of plasma is completely cleared of any substance by its passage through the kidney. Nevertheless, the clearance formula has considerable usefulness in renal physiology and for assessing renal function in disease. Below, we consider the clearances of two specific substances, inulin and *p*-aminohippuric acid, which can be used to measure the glomerular filtration rate and the renal plasma flow, respectively.

Before we look at the clearance of these substances in detail, an understanding of the technical requirements for accurate clearance measurements is necessary. Because the plasma concentration of the substance, x, must be known accurately, it must either be constant or changing in a predictable way so that an accurate average concentration can be calculated. So clearance measurements are only suitable for the steady-state determinations of GFR and RBF and cannot be used if rapid or transient changes are occurring. An additional complication is that urine flow must be adequate to collect sufficient for the assay in the clearance period (which is usually 10–20 min), so clearance measurements are not possible in conditions of anuria.

7.1.1 *Inulin clearance: the measurement of glomerular filtration rate*

Inulin is a polysaccharide with a molecular weight of approximately 5500. It is not a normal constituent of the body, but can be injected (or, usually, infused) intravenously in order to measure the inulin clearance.

Inulin is small enough to pass through the glomerular filter without difficulty, but is neither reabsorbed, secreted, synthesized nor metabolized by the kidney. So all the filtered inulin is excreted and all the inulin which is excreted has entered the urine only by filtration at the glomerulus.

The amount of inulin excreted per minute is $U_{in}V$, i.e. the urinary inulin concentration, U_{in} (mg/ml) multiplied by the urine flow, V (ml/min), and therefore this is the amount which entered the nephron by being contained in filtered plasma. The volume of plasma from which the amount $U_{in}V$ mg/min of inulin was derived must therefore have been

$$\frac{U_{in}V}{P_{in}}$$

and this is the clearance formula, where P_{in} is the plasma concentration of inulin (mg/ml).

Thus the inulin clearance, C_{in}, is equal to the glomerular filtration rate (GFR). The inulin clearance measurement (and hence the GFR measurement) is independent of the plasma inulin concentration, as the graph in Figure 7.1 shows.

The normal inulin clearance (GFR) is 125 ml/min (180 litres/day; it varies with body size; the value is really 125 ml/min per 1.73 m^2 body surface area). Even taking into account body surface area, GFR is low in infants and decreases in old age. From day to day the GFR is remarkably constant in man.

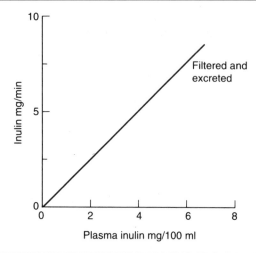

Figure 7.1   Inulin filtered and excreted as a function of plasma concentration. Since, after filtration, inulin is neither reabsorbed nor secreted, the amount excreted is identical to the amount filtered. The amount excreted (mg/min) is calculated as the product of the urinary inulin concentration U_{in}, and the urine flow, V. The inulin clearance, U_{in} V/P_{in}, is independent of the plasma inulin concentration (P_{in}), e.g. if P_{in} is 2 mg/100 ml (0.02 mg/ml) then $U_{in}V/P_{in}$, from the graph, is $2.5/0.02 = 125$ ml/min. If P_{in} is 4 mg/ 100 ml (0.04 mg/ml) then $U_{in}V/P_{in}$ is $5/0.04 = 125$ ml/min.

Variations in excretion of water and solutes depend on changes in tubular reabsorption and secretion, not on GFR changes.

Because inulin is not a normal constituent of the body, and measurements of inulin clearance therefore involve inulin infusions, it is rarely used clinically. A suitable alternative way to measure GFR is by using **creatinine clearance**. Creatinine is a product of muscle metabolism, so, as long as the subject remains at rest, the plasma creatinine level stays reasonably constant. Like inulin, it is freely filtered and not reabsorbed, synthesized or metabolized by the kidney. However, creatinine is secreted to some extent by the tubules, which makes the term UV artifactually high, but as the plasma creatinine assay is not absolutely specific and overestimates the true plasma creatinine concentration, the two errors tend to cancel each other and creatinine clearance is a reasonable estimate of GFR.

7.1.2   *Clearance ratios*

Since inulin is freely filtered, but is neither reabsorbed nor secreted, comparison of the clearance of any solute with the clearance of inulin can provide information about the net renal handling of the solute.

If a solute has a clearance value higher than the inulin clearance, then the solute must get into the renal tubules not only by glomerular filtration, but also by tubular secretion (PAH is such a substance, see below). However, if a solute has a clearance value less than that of inulin, there are two possibilities:

(a)   the solute may not be freely filtered at the glomerulus; or
(b)   the solute is freely filtered but is then reabsorbed from the tubule.

7.1.3   *PAH clearance: measurement of renal plasma flow*

PAH (*p*-aminohippuric acid) is one of the group of organic acids secreted by the proximal tubule. The secretory process is T_m-limited (see Chapter 5). PAH is not only secreted, but is also filtered at the glomerulus. Thus the amount excreted is equal to the sum of the amount filtered plus the amount secreted. This is illustrated graphically in Figure 7.2.

The T_m for PAH is reached at a plasma concentration of about 10 mg/100 ml plasma. At plasma concentrations below this, the clearance of PAH provides us with a measurement of the renal plasma flow (RPF). Why is this? In order to understand how PAH clearance can be a measure of RPF, it is necessary to introduce the **Fick** principle. The Fick principle is generally used to measure the lung blood flow – which is the same as the cardiac output. The formula is as follows

Lung blood flow (i.e. cardiac output)

$$= \frac{\text{Oxygen uptake/min}}{\text{Arteriovenous oxygen concentration difference}}$$

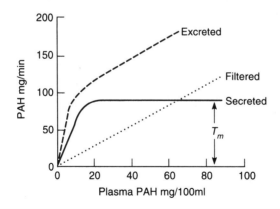

Figure 7.2   The relationship between plasma *p*-aminohippuric acid (PAH) concentration and the filtration, secretion and excretion of PAH (at a GFR of 125 ml/min). In addition to being filtered, PAH is also secreted, so the amount excreted/min is the sum of the amount filtered/min plus the amount secreted/min. The T_m for PAH secretion is reached at a rather low plasma PAH concentration (approximately 10 mg/100 ml).

Expressed in mathematical terms, this becomes

$$\text{Lung blood flow} = \frac{Q_{O_2}}{A_{O_2} - V_{O_2}}$$

where Q_{O_2} = ml O_2 used per min, A_{O_2} = arterial O_2 concentration, V_{O_2} = venous O_2 concentration. (The formula could also be applied using CO_2 output instead of O_2 uptake.)

The formula can be applied to any organ in which the blood takes up or loses any substance. So, in the kidney, let us suppose that the substance x is removed from the blood during its passage through the organ. Then:

$$\text{Renal blood flow} = \frac{\text{Amount } x \text{ excreted/min}}{\text{Arteriovenous concentration difference of } x}$$

$$= \frac{U_x V}{A_x - V_x}$$

where $U_x V$ is amount of x (mg) excreted per min, A_x is the renal arterial concentration of x, V_x is the renal venous concentration of x.

As long as the T_m is not exceeded, PAH is almost completely removed from the blood, and consequently the renal venous concentration can be considered to be zero. So applying the Fick formula to PAH, we can say

$$\text{Renal blood flow} = \frac{U_{PAH} V}{A_{PAH} - V_{PAH}}$$

but since V_{PAH} is zero,

$$\text{RBF} = \frac{U_{PAH} V}{A_{PAH}}$$

where A_{PAH} is, strictly, the renal arterial concentration of PAH. However, a venous concentration measurement can be used, provided that the venous sample does not include renal venous blood. This is usually the case because samples are generally taken from a limb, where the venous concentration is equal to the renal arterial concentration. In fact, since only plasma PAH is filtered, and we normally only measure PAH in the plasma (not whole blood), we measure the renal plasma flow (RPF) as

$$\text{Clearance of PAH} = \text{RPF} = \frac{U_{PAH} V}{P_{PAH}}$$

where P_{PAH} is the plasma PAH concentration. A typical figure for RPF obtained in this way is 600 ml/min.

A simpler way of understanding why PAH clearance is a measure of renal plasma flow is as follows. Because PAH is almost completely removed from the plasma by the kidneys (provided the T_m is not exceeded) and excreted in the urine, we can say:

PAH delivered to kidneys in plasma = PAH excreted in urine.

The PAH delivered to the kidneys is the renal plasma flow (RPF), multiplied by the plasma concentration of PAH (P_{PAH}). The urinary excretion of PAH is the urine flow (V) multiplied by the urinary concentration of PAH (U_{PAH}). So

$$RPF \cdot P_{PAH} = U_{PAH} V$$

therefore,

$$RPF = \frac{U_{PAH} V}{P_{PAH}}$$

and this is the clearance formula!

We can then use the haematocrit to obtain the renal blood flow, e.g. the usual haematocrit is 45%. This means that 45% of the total blood volume is cells and therefore 55% is plasma. Therefore

$$RBF = RPF \times \frac{100}{55} = 600 \times \frac{100}{55} = 1100 \, ml/min$$

The clearance of PAH, although generally used as a measure of renal plasma flow, does not in fact measure the plasma flow exactly. This is because PAH is not completely cleared from the blood during one passage through the kidney. The PAH extraction is about 90% complete. This incomplete removal is due to the fact that not all of the blood which enters the kidney goes to the glomeruli and tubules. Some goes to the capsule, the perirenal fat and the medulla (the blood in the vasa recta has had some PAH removed by filtration, but is not available for secretion in the proximal tubule). The PAH clearance approximates cortical plasma flow and is usually called the effective renal plasma flow (ERPF).

7.1.4 *Filtration fraction*

As inulin clearance is a measure of glomerular filtration rate, and PAH clearance is a measure of the effective renal plasma flow, simultaneous measurements of inulin and PAH clearance enable us to calculate the fraction of renal plasma that is filtered through the glomeruli into the nephrons.

$$\text{Filtration fraction} = \frac{C_{in}}{C_{PAH}} = \frac{125 \, ml/min}{600 \, ml/min} = 20\% \text{ in normal man}$$

7.1.5   *Other ways of measuring renal blood flow*

Clearance methods cannot tell us anything about the distribution of blood flow within the kidney. Several methods are available, however, which can provide information about the distribution of blood flow in experimental animals.

Inert gas 'washout' technique
This procedure is as follows: a small volume of saline containing radio-active krypton (^{84}Kr) or xenon (^{133}Xe) is administered by rapid injection into a renal artery (via a catheter). The lipid-soluble gas rapidly diffuses across the renal capillary membranes, so that the renal tissue becomes almost instantaneously saturated with the gas. The rate of removal ('washout') of the gas from tissue will then depend on the blood flow.

Isotope uptake technique
^{42}K or ^{86}Rb (radioactive potassium or rubidium) is administered by rapid intravenous injection and the animal is then killed. The rate of accumulation of the isotope in the different segments of renal tissue provides a measure of the blood flow through the regions.

7.2   *Regulation of renal blood flow and glomerular filtration rate*

Because the function of the renal blood supply is to provide blood for filtration, it is clear that an adequate supply of blood to the kidney is necessary if the normal excretory functions of the kidney are to continue. We have already seen that the renal blood supply is very large (about 1.1 l/min, or 20% of the cardiac output). An important feature of the renal blood flow is that, over a wide range of perfusion pressures (from a mean of 90 mmHg up to about 200 mmHg), the blood flow is independent of the perfusion pressure. This is true even if the kidney is denervated – i.e. it does not depend on the renal nerve supply. It also occurs in isolated perfused kidneys, so does not depend on blood-borne hormones. This property is therefore termed 'autoregulation' (Figure 7.3). The glomerular filtration rate also autoregulates (Figure 7.3). Essentially, the relationship between blood pressure and renal blood flow demonstrated in Figure 7.3 means that as the perfusion pressure increases, the resistance to flow also increases. Both afferent arterioles and efferent arterioles are capable of vasoconstriction. The effects are explained in Figure 7.4.

There is still some controversy about precisely how the autoregulation of renal blood flow and GFR occurs. The most widely accepted explanation is the **myogenic theory**, according to which the increase in wall tension of the afferent arterioles, brought about by an increase in perfusion pressure, causes

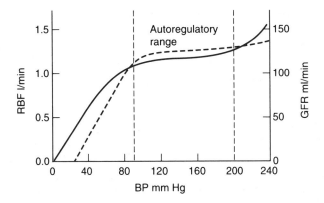

Figure 7.3 The autoregulation of glomerular filtration rate (...) and renal blood flow (—). In the autoregulatory range (90–200 mmHg), changes in mean arterial blood pressure have little effect on the renal blood flow (RBF) or the glomerular filtration rate (GFR).

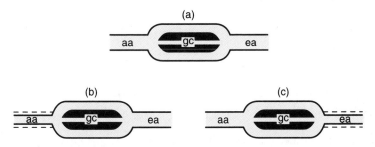

Figure 7.4 Diagram to show the ways in which vascular diameter can change in the afferent arterioles and efferent arterioles on either side of the glomerular capillary bed. (a) Normal situation, such as would occur at mean arterial blood pressure of about 100 mmHg, with GFR and RBF being maintained at their normal values (120 ml/min and 1100 ml/min respectively) by the combination of afferent and efferent arteriolar vasoconstrictor activity. (b) Afferent arteriolar constriction increases the overall renal vascular resistance and so reduces RBF. The afferent constriction also reduces the glomerular capillary hydrostatic pressure, hence decreasing the GFR. Afferent arteriolar dilatation (not shown) decreases the overall renal vascular resistance, so increasing the RBF. The glomerular capillary pressure hydrostatic pressure increases (since, in effect, more of the arterial pressure is transmitted through the afferent arteriole to the glomerular capillaries), and so the GFR increases.

(c) Efferent arteriolar constriction reduces the outflow from the glomerular capillaries into the efferent arteriole, and hence increases the glomerular capillary hydrostatic pressure and increases the GFR (i.e. a larger fraction of the plasma delivered to the kidney is filtered). Efferent arteriolar dilatation (not shown) increases the outflow from the glomerular capillaries into the efferent arterioles, so decreasing the glomerular capillary hydrostatic pressure and the GFR (so that a smaller fraction of the plasma delivered to the kidneys is filtered).

Changes in the diameter of the efferent arterioles have little effect on the overall renal vascular resistance (which is due mainly to the afferent arterioles) so are unlikely to change RBF.

automatic contraction of the smooth muscle fibres in the vessel wall, thereby increasing the resistance to flow and so keeping the flow constant in spite of the increase in perfusion pressure. There is considerable experimental evidence in support of this hypothesis, but the tubulo-glomerular feedback mechanism (p. 44) could also be involved.

7.2.1   *Further points concerning renal haemodynamics*

Because the kidneys exhibit the property of autoregulation it is easy to forget that renal haemodynamics do vary considerably. Autoregulation means that changes in blood pressure *per se*, in the autoregulatory range, have little effect on blood flow. It does not mean that blood flow is always constant. In many circumstances (e.g. physical or mental stress, haemorrhage), there are increases in the sympathetic nervous activity to the kidney (and to other parts of the body), causing vasoconstriction and hence a reduction in renal blood flow, even though the perfusion pressure is still in the autoregulatory range. *Hence the kidneys play a part in maintaining systemic blood pressure, rather than reacting passively to it.* However, renal vasoconstriction in response to renal sympathetic nerve activity is attenuated to some extent by the intrarenal production of vasodilator prostaglandins (p. 116). When the renal blood flow *is* reduced, there is generally an increased filtration fraction, brought about by efferent arteriolar vasoconstriction (mediated by angiotensin II (p. 112)), so that the GFR tends to be maintained.

The regulation of GFR may also involve other vasoactive agents, which are present within blood vessel walls throughout the body, but may be of particular importance in the kidney. Such agents include endothelin and nitric oxide. Endothelin is a peptide, present in the vascular endothelium, which is a potent vasoconstrictor of both afferent and efferent arterioles, and decreases GFR. Its production increases in many diseases in which renal function is impaired, including toxemia of pregnancy and diabetes mellitus; however, the extent to which endothelin is responsible for renal impairment in these circumstances is unclear. Nitric oxide is a vasodilator present in vascular endothelium, and it contributes to the maintenance of the normal vascular resistance in the kidney. Drugs which block nitric oxide synthesis constrict afferent (and efferent) arterioles, and decrease GFR.

Further reading

Lemley, K. V. and Baylis, C. (1997) The renal circulation. In R. L. Jamison and R. Wilkinson (eds), *Nephrology*. Chapman and Hall, London, pp. 34–43

Schuster, V. L. and Seldin, D. W. (1992) Renal clearance. In D. W. Seldin and G. Giebisch (eds), *The kidney, physiology and pathophysiology*, 2nd edition, Raven Press, New York, pp. 943–978

Age-related changes in GFR and RBF

In adults, the kidneys receive about 20% of the cardiac output. However, in the newborn infant, only about 5% of the cardiac output supplies the kidneys; this value increases progressively during the first year after birth.

The filtration of fluid, and consequent urine formation, begins at around the tenth week of gestation, and contributes to the amniotic fluid around the fetus.

Glomerular filtration rate increases at birth and immediately post-partum is about 25 ml/min per $1.73\,m^2$ body surface area (adult value, 125 ml/min per $1.73\,m^2$). From about one month there is a progressive increase in GFR, and the adult value, 125 ml/min per $1.73\,m^2$ is reached at about one year.

PAH clearance cannot be reliably used as a measure of renal plasma flow in young infants, since PAH is not completely cleared (i.e. the tubular secretion mechanism is not fully developed).

Visscher, C. A., Zeeuw, D. de, Navis, G., Zanten, A. K. van, Jong, P. E. de and Huisman, R. M. (1996) Renal 131-I hippurate clearance overestimates true renal blood flow in the instrumented conscious dog. *Am. J. Physiol.* **271**, F269–F274

Problem

Visscher, C. A., Zeeuw, D. de, Navis, G., Zanten, A. K. van, Jong, P. E. de and Huisman, R. M. (1996) Renal 131-I hippurate clearance overestimates true renal blood flow in the instrumented conscious dog. *Am. J. Physiol.* **271**, F269–F274

7.1 In an experiment on rats in which a PAH infusion was administered, the plasma PAH concentration was found to be 20 ug/ml and the urine PAH concentration was 1.333 mg/ml. The urine flow was 180 ul/min. The renal venous PAH concentration was also measured, and found to be 3 ug/ml.

 (a) What is the PAH clearance (ml/min)?

 (b) What is the effective renal plasma flow (ERPF)?

 (c) What is the total renal plasma flow?

 (d) Why are the effective renal plasma flow and total renal plasma flow measurements different?

Regulation of body fluid osmolality 8

8.1 *Introduction*

The body weight of a healthy adult on an adequate diet remains remarkably stable from day to day, and this stability indicates that the body fluid volume is staying constant, i.e. there is a steady state, in which the fluid output equals the fluid input. The normal intake and output of water over a 24-hour period is shown in Table 8.1.

It can be seen that water is lost from the body via several routes. The loss from the skin is 'insensible perspiration' and it occurs continually. Sweating (or 'sensible' perspiration) represents an additional loss (not shown in the table, but it can be up to 5 l/h). The loss from the respiratory tract occurs because inspired air is moistened as it passes to the lungs, and some of this moisture is then lost from the body in the expired air. This loss is greater in very dry environments, such as in deserts or in subzero temperatures. The faecal loss is normally small, but in pathological states faecal losses can lead to very severe dehydration – which may frequently be the cause of death, e.g. in cholera. This is because the gastrointestinal tract not only has to reabsorb the water taken in by mouth (1500 ml/day), but also must reabsorb the secreted

Table 8.1 Normal body intake and output of water over 24 hours

Intake (ml)		Output (ml)	
Drinking	1500	Urine	1500
Water in food	500	Respiration	400
Metabolism	400	Skin	400
		Faeces	100
Total	2400	Total	2400

96

digestive juices (up to 10 l/day). Failure to do so leads to very large water losses. Such losses can be more serious in young infants, and serious body fluid disturbances may occur as a result of diarrhoea.

From the foregoing, it is apparent that the fluid losses from the skin and respiratory tract, and in the faeces, are potential disturbing factors for body fluid balance. In contrast, the loss from the kidney can be regulated, so that the loss of fluid in the urine and the urine composition are adjusted to keep the body fluid volume and composition constant. (There are, of course, limits to the regulatory capacity of the kidney. If no water at all is ingested, then the urine becomes maximally concentrated and small in volume, but urinary water loss cannot be reduced below about 300 ml/day.)

The normal plasma osmolality (P_{osm}) is 280–290 mosmol/kg H_2O; it is regulated very precisely and a variation in either direction of about 3 mosmol/kg H_2O results in the operation of the body's osmolality regulating mechanisms.

8.2 *Osmoreceptors*

Alterations of the plasma osmolality are detected by **osmoreceptors** in the vicinity of the supraoptic and paraventricular areas of the anterior hypothalamus, which are supplied with blood by the internal carotid artery. These receptors regulate the release of the hormone ADH (antidiuretic hormone, vasopressin). These receptors, and others in the lateral preoptic area of the hypothalamus, also affect thirst.

The addition to the body fluid of excess water lowers the plasma osmolality, i.e. the solutes are diluted. Water deficiency, in contrast, increases the plasma osmolality. These changes lead to an appropriate change in ADH release (Figure 8.1).

The functional characteristics of the **osmoreceptors** can be seen in Figure 8.2; the relationship between the plasma osmolality and the plasma ADH concentration is such that at normal plasma osmolality (i.e. about 285 mosmol/kg H_2O) there is ADH present in the plasma; lowering the plasma osmolality reduces the ADH concentration and raising the plasma osmolality increases the plasma ADH concentration. The effect of changing plasma ADH concentrations on the urinary osmolality is shown in Figure 8.3. This coupling of the ADH-sensitive concentrating mechanism to the precise control of ADH release by osmoreceptors provides a very good regulatory mechanism for plasma osmolality.

8.2.1 *Sensitivity of the osmoreceptors to osmotic changes caused by different solutes*

Normally, sodium and its associated anions contribute over 95% of the osmotically active constituents of the plasma, and where osmotic changes are

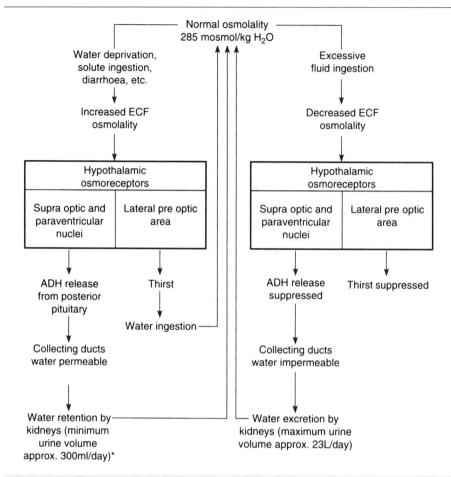

Figure 8.1   The regulation of osmolality by ADH. Note: a volume as low as 300 ml would only be possible if solute output were very low. (See Figure 8.6) The minimum possible urine volume will depend on the amount of solute which has to be excreted, since the maximum urine concentration is about 1400 mosmol/kg H_2O.

brought about by loss or gain of water in the body, the Na^+ concentration will be altered. However, addition or loss of solutes unaccompanied by water also changes the plasma osmolality (as discussed in more detail in the following chapter) and not all solutes are equally effective osmoreceptor stimulants. Their effectiveness depends on the degree to which they are unable to cross cell membranes (i.e. their ability to cause cellular dehydration). Thus sodium ion is an effective osmoreceptor stimulus, but urea or potassium are much less effective.

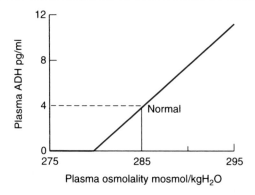

Figure 8.2 ADH release from the posterior pituitary alters when plasma osmolality changes. If the plasma osmolality increases, ADH release increases to raise the plasma ADH level. Decreases of plasma osmolality reduce the plasma ADH level.

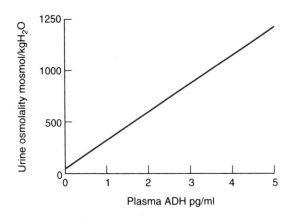

Figure 8.3 The relationship of plasma ADH concentration to urine osmolality. A tenfold change in plasma ADH concentration (from 0.5 to 5 pg/ml) is responsible for almost the full range of urine osmolalities.

8.2.2 *Synthesis and storage of ADH*

Antidiuretic hormone is synthesized in the hypothalamus (supraoptic nucleus) as part of a large precursor molecule (166 amino acids) shown diagrammatically in Figure 8.4. The precursor molecule, after synthesis, is transported from the hypothalamus to the neurohypophysis (posterior pituitary) within nerve fibres which constitute the hypothalamohypophyseal tract, and is thought to be cleaved progressively as it moves down the axons. In the nerve terminals, ADH and neurophysin are stored as insoluble complexes in

Figure 8.4 Schematic diagram of the ADH precursor molecule. The molecule is 166 amino acids, and can be divided into four segments, a leader sequence, the ADH moiety, neurophysin, and a 39 amino acid glycopeptide.

secretory granules. Upon release, these complexes dissociate rapidly, due to dilution and the increase in pH (the storage granules are acidic). The release of ADH (and neurophysin) occurs in response to action potentials in the neurones from the hypothalamus, which contain ADH. The action potentials depolarize the cell membrane, leading to calcium influx, which triggers fusion of the secretory granules with the cell membrane and release of their contents.

8.2.3 *Structure of ADH*

Antidiuretic hormone is a nonapeptide if the sulphur-containing amino acids are regarded as two linked cysteine residues (or an octapeptide if they are regarded as a single cystine residue). The molecular weight is just over 1000 and the structure of human ADH is

$$\text{Cys-Tyr-Phe-Gln-Asn-Cys-Pro-Arg-Gly(NH}_2)$$

$$\lfloor\_S \underline{\hspace{3cm}} S\_\rfloor$$

This is generally termed 8-arginine-vasopressin, to distinguish it from the antidiuretic hormone found in pigs and some other species, where the arginine residue is replaced by lysine to form 8-lysine-vasopressin.

8.2.4 *Cellular actions of ADH on water permeability*

ADH only increases water permeability when it is present on the peritubular side of the collecting tubule cell, and is ineffective from the luminal side. This is because the renal ADH receptors (V_2 receptors) are on the basal membranes of the tubule cells.

The V_2 receptor is a member of the G-protein coupled receptor group, characterized by seven membrane-spanning regions. Binding of ADH to the receptor activates the enzyme adenylate cyclase, which catalyses the formation of cyclic 3′, 5′-AMP (adenosine monophosphate) from ATP. The cyclic AMP thus generated activates a further enzyme, protein kinase, which phosphorylates proteins in non-operative water channels located close to the apical membrane and hence increases the incorporation of the water channels (aquaporin-2) into the apical membrane to become functional (Figure 8.5).

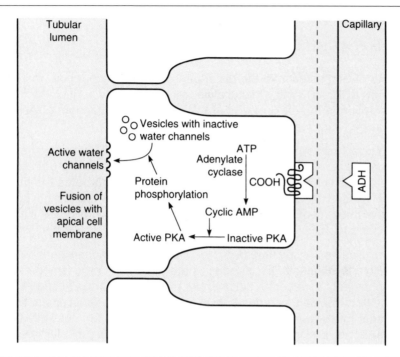

Figure 8.5 The cellular actions of ADH in the collecting tubule (principal cells) leading to the incorporation of water channels in the apical cell membrane and hence an increase in water permeability. PKA = protein kinase A.

8.2.5 *Removal of ADH from the blood*

In order for ADH to regulate the plasma osmolality precisely, it is not only necessary that ADH should be rapidly released in response to dehydration and that the release is rapidly stopped when the dehydration is corrected, but also that ADH must be rapidly removed from the plasma. This removal occurs in the liver and in the kidney. Of the ADH removed from the plasma by the kidneys (about 50% of the total), less than 10% appears in the urine, the remainder being metabolized.

The plasma half-life of ADH (i.e. the time taken for the plasma ADH concentration to fall by 50%) is 10–15 min.

8.2.6 *Drugs affecting ADH release*

A number of pharmacological agents may alter ADH release and thereby disturb osmoregulation. Probably the most widely used drug which increases

---

Diabetes insipidus (DI)

The water reabsorption in the distal part of the nephron, under the influence of ADH, is termed 'facultative' reabsorption, whereas the water reabsorption in the earlier parts of the nephron, independent of ADH, is 'obligatory' water reabsorption.

Facultative reabsorption (ADH-dependent) is about 23 litres/day, i.e. this volume can be either reabsorbed or excreted. Some people are unable to reabsorb this volume and have the condition termed **diabetes insipidus**. It may be caused by a congenital absence of ADH production by the hypothalamus (central diabetes insipidus), or by a failure of the kidneys to respond to ADH (nephrogenic diabetes insipidus).

Central diabetes insipidus can be readily treated by administering ADH. Usually a synthetic ADH is administered (DDAVP or desmopressin), which has only V_2 receptor actions, not V_1 (vasoconstrictor) actions. This can be administered as a nasal spray without causing vasoconstriction in the nasal mucosa. Nephrogenic diabetes insipidus is not currently treatable. It is not a single condition. One form of nephrogenic DI is an X-linked congenital defect in the V_2 receptor. Another form, which is autosomal and recessive, is a mutation of the aquaporin-2 molecule which renders it non-functional.

ADH release is nicotine. Other drugs with a similar effect include ether, morphine and barbiturates. Conversely, some drugs, notably alcohol, inhibit ADH release.

8.3 *Regulation of water excretion and water reabsorption*

When the body fluids have become hypertonic (e.g. because of dehydration), the function of the kidneys is to reabsorb 'pure' (i.e. osmotically 'free') water from the tubular fluid, to dilute the plasma. In the process, the urine becomes concentrated. Conversely, if excess water has been ingested, the function of the kidney is to excrete this excess 'pure' (osmotically 'free') water, so that dilute urine is produced. Thus, in the production of concentrated urine, osmotically free water is reabsorbed, whereas in the production of dilute urine, osmotically free water is excreted. (Note that 'dilute' urine has a lower osmolality than plasma, and 'concentrated' urine has a higher osmolality than plasma.)

Whether the kidneys produce dilute or concentrated urine depends primarily on the level of circulating ADH, since changes in ADH levels affect not only the volume of urine excreted, but also the urinary concentration. This is because, although ADH has little effect on the quantity of solute excretion, the volume of water in which the solutes are excreted is altered by ADH.

If the kidney is producing urine which is isotonic to the plasma, then the osmotically active constituents of the urine are being excreted in a volume of water sufficient to keep the solutes at the same osmotic pressure as the plasma. This volume of water, in ml/min, is the rate at which osmotically active substances are cleared from the plasma, i.e. it is the osmolar clearance. We can define osmolar clearance using the clearance formula (p. 86) as

$$\text{Osmolar clearance, } C_{osm} = \frac{U_{osm} V}{P_{osm}}$$

where U_{osm} and P_{osm} are, respectively, the urine and plasma osmolalities, and V is the urine flow (ml/min). In the case cited above, of the urine being isotonic to the plasma, then

$$\frac{U_{osm}}{P_{osm}} = 1$$

so, that

$$C_{osm} = V$$

If the urine osmolality is lower than the plasma osmolality, i.e. if dilute urine is being produced, then, since U_{osm}/P_{osm} is less than 1, C_{osm} must be less than V. Another way of saying this is that the urine volume per minute, V, is made up of an additional volume of 'free' water (C_{H_2O}) as well as isotonic fluid (C_{osm}). So that:

$$V = C_{osm} + C_{H_2O}$$

where C_{H_2O} is the free water clearance. Free water is excreted when ADH levels are low (so that the urine is dilute); the excretion of free water raises the plasma osmolality, P_{osm}. The excretion of free water depends on (a) the generation of osmotically free water in the ascending limb of the loop of Henle by NaCl absorption unaccompanied by water, and (b) the excretion of this osmotically free water because of the impermeability of the collecting ducts, preventing its reabsorption. The maximum C_{H_2O} in man is 12–15 ml/min (15–22 litres/day).

If the urine osmolality is greater than the plasma osmolality (i.e. if concentrated urine is being produced), then U_{osm}/P_{osm} is greater than 1, and V must be less than C_{osm}. So in the equation

$$V = C_{osm} + C_{H_2O}$$

C_{H_2O} must be negative. However, it is rather confusing to speak of negative free water clearance: what this really means is that free water is being absorbed, not excreted. So we can introduce the term **free water reabsorption**, $T^c_{H_2O}$ which is the same as $-C_{H_2O}$

$$V = C_{osm} - T^c_{H_2O}$$

so that

$$T^c_{H_2O} = C_{osm} - V$$

For example, suppose that $V = 1$ ml/min, $P_{osm} = 290$ mosmol/kg H_2O and $U_{osm} - 1000$ mosmol/kg H_2O. Then

$$C_{osm} = \frac{U_{osm}V}{P_{osm}} = \frac{1000}{290} \times 1 = 3.45\,\text{ml/min}$$

and

$$T^c_{H_2O} = C_{osm} - V$$
$$= 3.45 - 1$$
$$= 2.45\,\text{ml/min}$$

This 2.45 ml/min is the volume of solute-free water being returned to the plasma by renal absorption.

In circumstances in which there is solute-free water reabsorption, free water is still generated in the ascending limb of the loop of Henle, but this (and more) is reabsorbed during the passage of the tubular fluid along the collecting tubules, so that the final urine is more concentrated than the plasma.

Measurements of C_{H_2O} and $T^c_{H_2O}$ are quantitative ways of determining the ability of the kidney to excrete or conserve water. They are also useful measurements to indicate the physiological sites of action of diuretics (see Chapter 15).

8.3.1 *Effect of solute output on urine volume*

Because the concentrating ability of the kidneys is limited (maximum urinary osmolality is about 1400 mosmol/kg H_2O), it follows that the amount of urine excreted per day can depend not only on the level of circulating ADH, but also on the amount of solute to be excreted (Figure 8.6).

For example, suppose we have 800 mosmol of solute to excrete per day. Since the maximum urinary osmolality is 1400 mosmol/kg H_2O, the minimum

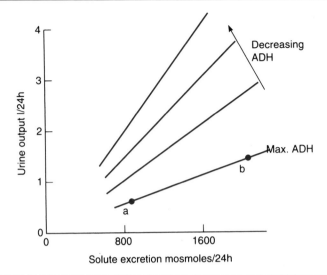

Figure 8.6 The effect of ADH on urine output. Since ADH enables urine to be concentrated by water abstraction from the collecting ducts, the urine in the presence of maximal ADH has an osmolality of about 1400 mosmol/kg H_2O. However, the amount of urine produced per day, of this osmolality, will depend on the amount of solute to be excreted. Point (a) shows the minimum possible urine output with a solute excretion of 800 mosmol/24 h, whereas point (b) shows the minimum urine output if 2000 mosmol/24 h have to be excreted.

volume of urine which can be excreted is

$$\frac{800}{1400} \text{ kg } H_2O/24\,h = 571\,ml/24\,h$$

Shown as point (a) in Figure 8.6.

If the amount of solute to be excreted per day were 2000 mosmol, then the minimum volume of urine which can be excreted is

$$\frac{2000}{1400} \text{ kg } H_2O/24\,h = 1430\,ml/24\,h$$

Shown as point (b) in Figure 8.6. Thus the excretion of large amounts of solute causes a diuresis, even in the presence of high levels of ADH.

From the foregoing, it should be apparent why it is not possible for man to survive by ingesting very hypertonic solutions. Suppose that a subject ingests 1 litre of 2000 mosmol/kg H_2O solution. Since the maximum urine osmolality is 1400 mosmol/kg H_2O, the excretion of this solute cannot occur in less than

$$\frac{2000}{1400} \text{ kg } H_2O = 1430\,ml$$

Age-related changes in urinary concentrating and diluting ability

Concentrating ability

The maximum urinary concentration attainable by newborn infants is only about 600 mosmol/kg H_2O (i.e. about one-half the adult value). The concentrating ability increases to approach the adult value by the end of the first year. The poor concentrating ability is related to low ADH levels and to a poorly developed cortico-medullary osmotic gradient (partly due to low urea excretion, due to the highly anabolic state of infants).

The poor concentrating ability of the infant kidney has important consequences. A typical solute load to be excreted by a 3-month-old infant is 150 mosmol per 24 hours. With a maximum concentrating ability of 600 mosmol/kg H_2O, this solute output will require a minimum urine volume of

$$\frac{150}{600} = 0.25 \, kg \, H_2O \, per \, 24 \, h = 250 \, ml.$$

This is about one-half the water intake of a child of this age, so modest reductions in fluid ingestion, or increases in solute ingestion, can lead to dehydration.

Concentrating ability decreases in old age (60 years and over). Typically, the maximum urine concentration of a 70-year-old is about 900 mosmol/kg H_2O.

Diluting ability

In contrast to the concentrating ability, the diluting capability of the infant kidney is well-developed. Newborn infants have a minimum urine concentration of about 35 mosmol/kg H_2O (circa 60 mosmol/kg H_2O in adults) and this enhanced diluting capacity lasts until about one year of age.

Thus more water would be required to excrete the solute than was ingested with it and the subject becomes dehydrated.

Some non-reabsorbable solutes, such as mannitol, impair the renal concentrating ability and lead to the production of an almost isotonic urine. Such solutes are termed 'osmotic diuretics' and are fully discussed in Chapter 15.

8.3.2 *Adrenal steroids and the renal response to ADH*

Adrenal insufficiency reduces the glomerular filtration rate, and this impairment is remedied by the administration of glucocorticoids. Adrenal insufficiency also impairs both the renal concentrating ability, and the renal response to water loading, i.e. very dilute urine cannot be produced. In the absence of adrenal steroids (glucocorticoids), the collecting duct water permeability is above the basal level, even in the absence of ADH, and the effect of ADH on permeability appears to be reduced.

Further reading

Berl, T. (1998) Water channels in health and disease. *Kidney Int.* **53**, 1417–1418

Burg, M. B. (1995) Molecular basis of osmotic regulation. *Am. J. Physiol.* **268**, F983–F996

Garcia, N. H., Pomposiello, S. I. and Garvin, J. L. (1996) Nitric oxide inhibits ADH-stimulated osmotic water permeability in cortical collecting ducts. *Am. J. Physiol.* **270**, F206–F210

King, L. S. and Agre, P. (1996) Pathophysiology of the aquaporin water channels. *Annu. Rev. Physiol.* **58**, 619–648

Regulation of body fluid volume

9

9.1 Introduction

Sodium salts are the main osmotically active solutes of the extracellular fluid (ECF). Thus, since body fluid osmolality is regulated by osmoreceptors and ADH release, if the extracellular fluid sodium *content* changes, then ECF volume will also change (Figure 9.1).

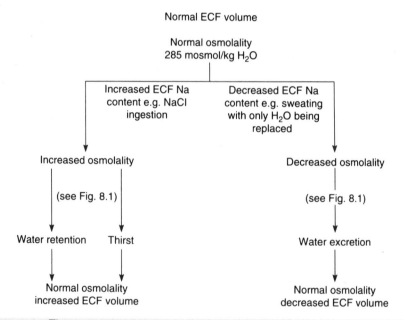

Normal ECF volume

Normal osmolality
285 mosmol/kg H_2O

Increased ECF Na content e.g. NaCl ingestion

Decreased ECF Na content e.g. sweating with only H_2O being replaced

Increased osmolality

Decreased osmolality

(see Fig. 8.1)

(see Fig. 8.1)

Water retention   Thirst

Water excretion

Normal osmolality increased ECF volume

Normal osmolality decreased ECF volume

Figure 9.1 The presence of an osmoregulatory mechanism means that changes in ECF sodium content will change the ECF volume (so that ECF sodium concentration stays constant).

It follows that the regulation of ECF volume (and, indirectly, of total body fluid) is brought about by the regulation of the body sodium content. This control of body sodium content is a vital function of the kidneys. In order to understand how this control is effected, we must introduce the concept of 'effective circulating volume'.

9.1.1 *'Effective circulating volume'*

Changes in the extracellular fluid Na^+ content, because they change the ECF volume, will affect the volume of blood perfusing the tissues. The 'effective circulating volume' is the component of blood which is perfusing the tissues. It should be noted that the effective circulating volume is not necessarily identical to the intravascular (blood) volume, since there may be circumstances (e.g. heart failure) in which not all of the blood is effectively perfusing the tissues (see Chapter 14).

When osmotic changes occur in the body, the response (increased or decreased ADH release) takes place within minutes, and the disturbance is rapidly corrected. In contrast, disturbances in body sodium content (and, consequently, in body fluid volume) may take many hours, or even days, to correct.

The maintenance of the body fluid volume depends on an adequate sodium intake. In the dog, if sodium is completely excluded from the diet, losses from the body (via the skin and kidneys) occur at the rate of about 25 mmol Na^+ per day. And, because osmoregulation occurs, there is a gradual reduction in the body fluid volume (this reduction is primarily of ECF volume). Man is able to conserve sodium more effectively than the dog and the urine can be almost free of sodium (urinary Na^+ can be less than 1 mmol/l). However, this maximal sodium reabsorption can affect the excretion of K^+ and H^+, and so may disturb acid–base balance (see Chapter 10).

Sodium excretion can be altered either by changes in filtration or by changes in reabsorption. Changes in Na^+ excretion are normally brought about by changes in tubular reabsorption. However, sufficiently large changes in ECF volume can also change GFR.

Exactly how Na^+ reabsorption is modified by changes in ECF volume (and effective circulating volume) is still a matter of debate. Nevertheless, many facts are clear and we can therefore attempt to integrate these facts in a coherent story of how Na^+ reabsorption is regulated.

9.2 *Aldosterone*

If the adrenal glands are removed, a number of metabolic defects appear and, in most species, death occurs within about two weeks. The cause of death is generally the loss of NaCl from the body, leading to circulatory collapse. There is excessive NaCl (and very little K^+) in the urine and the body sodium content falls (whereas the potassium content increases). These changes lead to the movement of K^+ (and water) into the cells so that the intracellular fluid volume increases. The ECF volume is therefore markedly decreased because of NaCl and water loss in the urine and the movement of water into the cells.

The above changes are due to the absence of the adrenal cortical hormone, **aldosterone**. Death can usually be postponed, after removal of the adrenal glands, either by a high sodium diet (to maintain ECF volume in spite of the urinary loss of NaCl and water), or by the administration of aldosterone.

It is thus apparent that aldosterone is necessary for normal sodium reabsorption to occur. This does not necessarily mean that aldosterone regulates Na^+ excretion; only that aldosterone is necessary for such regulation to occur, i.e. aldosterone has a permissive action, permitting Na^+ regulation. This caution in interpreting the role of aldosterone in Na^+ regulation is necessary because, when there are abnormalities in the secretion of aldosterone, Na^+ reabsorption and excretion, and thus regulation of ECF volume, are often unimpaired. This is because, although aldosterone does have a regulatory role, other mechanisms can compensate for abnormalities in aldosterone release.

Aldosterone stimulates Na^+ reabsorption from the distal nephron (primarily the cortical collecting tubule (p. 84)). Aldosterone also promotes H^+/K^+ secretion (p. 83). In addition to its renal actions, it also promotes Na^+ reabsorption from the colon and gastric glands, and from the ducts of sweat glands and salivary glands.

Changes in plasma aldosterone levels are usually accompanied by parallel changes in ADH levels. This is because hypovolaemia acts as a stimulus for the release of both hormones (see also p. 120). Together, the two hormones (aldosterone and ADH) can be considered as leading to the reabsorption of an isotonic absorbate which maintains the ECF volume.

However, when excess aldosterone is produced, or if aldosterone is administered, in excess, sodium retention (and oedema) rarely occurs – i.e. there is an 'escape' mechanism, whereby sodium excretion returns to normal in spite of the abnormal aldosterone level. This phenomenon emphasizes the complicated nature of body fluid volume regulation, and the fact that aldosterone, although it is *one of* the regulators of sodium excretion, cannot be said to be *the* regulator of sodium excretion.

9.2.1   *Control of aldosterone release*

Plasma K^+ concentration

Increases in plasma potassium concentration have a direct action on the adrenal cortex, stimulating aldosterone secretion. Very small changes in the plasma K^+ concentration (e.g. of 0.1 mM) can significantly increase aldosterone release. The consequence of this is increased distal tubular K^+ secretion, returning the plasma K^+ concentration to normal, so that aldosterone release decreases.

Plasma Na$^+$ concentration
Decreases in plasma sodium concentration have a direct effect on the adrenal cortex, to stimulate aldosterone secretion. However, this is not an important way in which aldosterone release is altered, because osmoregulation normally keeps the plasma sodium concentration constant.

Changes in the extracellular fluid volume
Because of the osmoregulatory mechanism, decreases in the body sodium content lead to decreases in the effective circulating volume. Changes in the effective circulating volume alter aldosterone release via the renin–angiotensin system.

9.3   *Renin and angiotensin, and their relationship to aldosterone*

Renin, the enzyme synthesized and stored in the juxtaglomerular apparatus, is released into the plasma when the body sodium content decreases. The body sodium content does not directly provide the stimulus for renin release; the stimuli are some of the physiological changes brought about by altered body sodium content. As noted earlier, body sodium content determines the effective circulating volume and it is this relationship that determines renin release.

There are three main ways in which decreases in effective circulating volume elicit increases in renin release.

(1)   Increased sympathetic nerve activity: if there is a reduction of the extra-cellular fluid volume and/or the effective circulating volume (which will tend to decrease the systemic arterial blood pressure), baroreceptor reflexes lead to increased sympathetic activity to arterioles. The main baroreceptors are in the carotid arteries (carotid sinuses). The renin-containing granular cells of the renal afferent arterioles are innervated by sympathetic nerve fibres; activity in these causes renin release, and is increased by decreases in systemic blood pressure. This nervous control of renin release is mediated by β-adrenergic receptors (also activated by circulating catecholamines) and is important in maintaining renin release under basal conditions and when assuming the upright posture, as well as during sodium depletion.

(2)   Decreased wall tension in the afferent arterioles: decreased renal perfusion pressure (such as will tend to occur if the effective circulating volume is reduced) leads to increased renin release from the granular cells. The immediate stimulus appears to be the wall tension (and its rate of change, i.e. the pulse pressure) at the granular cells, and hence this mechanism is termed the afferent arteriolar baroreceptor. Since the afferent arteriole is constricted by renal sympathetic nerve activity and this constriction is

'upstream' from the granular cells, sympathetic vasoconstriction of the kidney will increase renin release by the afferent arteriolar baroreceptor mechanism, in addition to there being direct release of renin by the receptor mechanism mentioned above.

(3) The macula densa mechanism: a decrease in the delivery of NaCl to the macula densa leads to renin release, but many of the details of this mechanism are still controversial. It is not clear whether it is Na^+ or Cl^- which is sensed (or both, or osmolality), or whether tubular NaCl concentration, or content (i.e. depending on concentration and tubular flow) or rate of NaCl transport is detected. A fall in NaCl delivery to the macula densa will occur in response to decreases in effective circulating volume, because such decreases in volume lead to avid NaCl reabsorption in the proximal tubules (the mechanism is described on p. 114), and may also lead to a fall in GFR.

The release of renin by the macula densa mechanism is mediated by renal cortical prostaglandins (mainly PGI_2; p. 116), i.e. the macula densa stimulus releases PGI_2 which acts on the granular cells to release renin. The release of renin by the renal nerves (β-adrenergic receptor mechanism) is independent of renal prostaglandin synthesis and the involvement of prostaglandins in afferent arteriolar baroreceptor-mediated renin release is unlikely.

Following stimulation of the juxtaglomerular apparatus, renin is released into the blood, where it acts on a plasma protein (an α_2-globulin), angiotensinogen (also called 'renin substrate') and splits off a decapeptide, angiotensin I. An enzyme on the surface of endothelial cells, 'converting enzyme', rapidly removes a further two amino acids from angiotensin I, to form the octapeptide, angiotensin II (Figure 9.2).

9.3.1 *Actions of angiotensin II (Figure 9.3)*

Angiotensin II acts on the zona glomerulosa of the adrenal cortex to release aldosterone. Aldosterone (Figure 15.1), in common with all adrenal cortical hormones, is synthesized from cholesterol (which itself may be synthesized within the gland or taken up from the circulation). Very little aldosterone is stored. Stimuli to its release promote aldosterone biosynthesis.

In spite of the connection between renin release, angiotensin II formation and aldosterone release, it is apparent from the 'escape' phenomenon (p. 108) that the actions of angiotensin II which are independent of its effect on aldosterone play an important part in the regulation of sodium excretion. Indeed, *angiotensin II can be regarded as the primary hormone in sodium regulation.* Angiotensin II is an extremely potent vasoconstrictor; this action plays a (normally minor) part in the maintenance of systemic blood pressure, but could also have important intrarenal consequences, perhaps by altering the

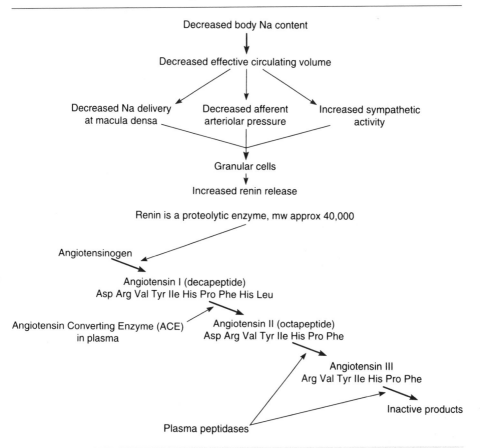

Figure 9.2 The regulation of angiotensin formation by body sodium content. Renin, released in response to decreased body Na^+ content, acts on a plasma α_2-globulin, angiotensinogen (renin substrate), to release angiotensin I. Angiotensin-converting enzyme in plasma converts this into angiotensin II, which has the actions shown in Figure 9.3. Angiotensin II is degraded to inactive products. In some species (but not man), angiotensin III (septuapeptide), like angiotensin II, can release aldosterone from the adrenal cortex.

distribution of glomerular filtration. Angiotensin II elicits vasoconstriction in efferent arterioles at lower concentrations than those needed to vasoconstrict afferent arterioles, so that at low (i.e. physiological) concentrations it does not decrease GFR, and indeed plays a part in maintaining GFR.

A third important role of angiotensin II is its direct effect on proximal tubular sodium reabsorption; physiological concentrations of angiotensin II increase proximal tubular Na^+ reabsorption. Both the apical and the basal membranes of proximal tubule cells have angiotensin II receptors; the main

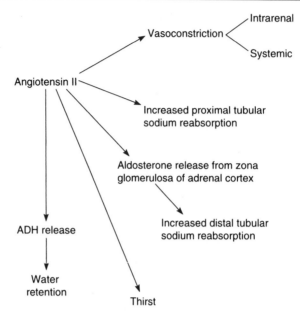

Figure 9.3   Actions of angiotensin II. For relative importance of these actions, see text.

way in which angiotensin II increases proximal sodium reabsorption is by increasing apical membrane Na^+/H^+ exchange.

Other actions of angiotensin II are the release of ADH and the stimulation of the sensation of thirst, by an action on the brain. Angiotensin II also has an inhibitory effect (negative feedback) on renin release by the juxtaglomerular granular cells.

9.4 *Starling forces and proximal tubular sodium reabsorption*

When the body sodium content changes, the modifications of effective circulating volume that result automatically adjust proximal tubular sodium reabsorption in the direction required to correct the disturbance. We saw in Chapter 5 (Section 5.3) that the uptake of sodium chloride from the lateral intercellular spaces of the proximal tubule into the capillaries depends on Starling forces. These forces therefore determine net sodium chloride reabsorption in the proximal tubule.

The rate of uptake of sodium chloride and water from the proximal tubule into the peritubular capillaries is determined by the rate of uptake from the lateral intercellular spaces into the capillaries; this rate of uptake is shown by the equation:

$$\text{capillary uptake}$$
$$\propto \text{forces favouring uptake} - \text{forces opposing uptake}$$
$$\propto (\Pi_{\text{cap}} + P_{\text{LIS}}) - (\Pi_{\text{LIS}} + P_{\text{cap}})$$

where Π is oncotic pressure, P is hydrostatic pressure, LIS is lateral intercellular space, and cap is capillary.

Both the peritubular capillary hydrostatic pressure and the plasma protein osmotic pressure (oncotic pressure) are altered by changes in effective circulating volume. Increases in volume (caused, for example, by increased NaCl and water content of the body), result in increased peritubular capillary hydrostatic pressure and, at the same time, dilute the plasma proteins and hence reduce the oncotic pressure. From the equation above, it is clear that these changes will decrease capillary uptake, and so decrease proximal tubular reabsorption of NaCl and water, thereby increasing the delivery of NaCl and water to more distal parts of the nephron.

The increased peritubular capillary hydrostatic pressure in such circumstances is due to an increased **venous pressure** (caused by increased circulating volume), rather than the transmission of an increased arterial pressure to the capillaries.

Decreases in effective circulating volume will tend to have the opposite effect – i.e. proximal NaCl and water absorption will be enhanced. The consequent decrease in the NaCl and water delivery to the macula densa in the distal nephron will act as a stimulus to renin release.

Although proximal tubular NaCl and water reabsorption are automatically adjusted to the requirements of the body for sodium excretion or sodium conservation, changes in proximal sodium reabsorption are not necessarily reflected in the final urine. This is because NaCl reabsorption in the ascending limbs of long loops of Henle is directly dependent on NaCl delivery. Reduced proximal NaCl reabsorption will increase the delivery of NaCl to the ascending limbs, where more NaCl will be extruded into the medullary interstitium. So it might appear that the loops of Henle are 'sabotaging' the efforts of the proximal tubules to correct volume disturbances. However, the nephrons with short loops of Henle may not transport a sufficient amount of NaCl out of the ascending limbs to reabsorb all of the excess if the delivery from the proximal tubule increases. We could thus term the short-looped nephrons 'sodium-losing nephrons' and the long-looped nephrons 'sodium-retaining nephrons'. The final urinary composition of the short-looped nephrons may depend on proximal reabsorption to a much greater extent than that of the long-looped nephrons. Redistribution of the renal blood flow between the two populations of nephrons could play a part in the response to changes in effective circulating volume.

9.5   *Renal nerves*

The activity in the renal sympathetic nerves is regulated by arterial baroreceptors. Increased body fluid volume will tend to increase blood pressure, causing a reflex response mediated by the baroreceptor mechanism:

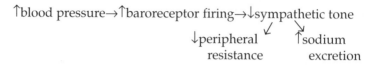

$\uparrow$blood pressure$\rightarrow\uparrow$baroreceptor firing$\rightarrow\downarrow$sympathetic tone

$\downarrow$peripheral $\uparrow$sodium
resistance   excretion

Conversely, decreases in body fluid volume, because they decrease the effective circulating volume, will tend to decrease blood pressure, which will reflexly (via the baroreceptors) increase sympathetic tone.

How do changes in the activity in the renal sympathetic nerves affect sodium excretion? The sympathetic nerves to the kidney go primarily to the afferent arterioles and mild stimulation of the renal nerves reduces the blood flow to the superficial nephrons and increases blood flow to the juxtamedullary nephrons, which, as we saw in the previous section, could lead to sodium conservation. This effect of the renal nerves may be mediated by angiotensin II, because renal nerve stimulation is known to release renin. This will also cause sodium retention by leading to aldosterone release, but, in addition, there is some evidence that catecholamines released from the renal sympathetic nerve endings (and from the adrenal medulla) stimulate sodium reabsorption by the proximal tubule, although whether this is a direct action on sodium transport or is due to altered peritubular forces, is a matter of conjecture at present. There is some evidence that the action is a direct one, since it has been demonstrated in isolated perfused renal tubules.

The renal nerves are not essential. Transplanted kidneys are of course denervated at the time they are transplanted, but they are nevertheless able to regulate body fluid volume and sodium excretion in an essentially normal manner.

9.6   *Prostaglandins and other arachidonate metabolites*

Prostaglandins are complex lipid molecules, which are synthesized in most cells of the body. In the kidney, there are at least three distinct sites of synthesis: (1) the cortex (including arterioles and glomeruli); (2) medullary interstitial cells; and (3) the collecting duct epithelial cells. The main prostaglandins synthesized by the kidney are PGE_2, PGI_2, $PGF_{2\alpha}$, PGD_2 and thromboxane A_2.

The renal prostaglandins are synthesized from arachidonic acid, which is stored esterified in membrane phospholipids. The release of free arachidonic acid occurs enzymatically and can be inhibited by steroids. Free arachidonic acid is converted by a cyclo-oxygenase enzyme (which can be inhibited by non-steroidal anti-inflammatory drugs) into unstable endoperoxides, which then give rise to the renal prostanoids. The major arachidonate products are, in the cortex, PGI_2 (prostacyclin) and, in the medulla, PGE_2.

The renal prostaglandins do not function as circulating hormones systemically since most are effectively catabolized by the lungs. However, the renal prostaglandins have important intrarenal effects. PGE_2 and PGI_2 are vasodilators, which are produced in increased amounts when renal perfusion is threatened (by vasoconstrictors such as noradrenaline (norepinephrine), vasopressin and angiotensin II), and thus minimize renal vasoconstriction. The inhibition of renal prostaglandin synthesis (e.g. by using non-steroidal anti-inflammatory drugs such as aspirin or indomethacin), has no effect on renal blood flow or GFR in normal individuals, but can lead to large falls in GFR and renal blood flow in subjects who have vasoconstrictor stimuli to the kidney (such as would occur when the effective circulating volume is decreased).

The anti-vasoconstrictor role of the renal prostaglandins is a cortical effect (PGI_2 is likely to be the most important prostanoid in this role). Hence *decreased* effective circulating volume leads to *increased* cortical prostaglandin synthesis. In this context, it should be remembered that PGI_2 is a mediator of renin release (p. 112).

The medullary prostaglandins (mainly PGE_2) have actions on the renal tubules (predominantly the collecting tubules) and are natriuretic and diuretic. They also impair the antidiuretic action of ADH. These actions may limit the extent to which Na^+ reabsorption (ATP-dependent) can be stimulated (e.g. by aldosterone) in the renal medulla, and hence protect the medullary tubule cells from excessive hypoxia during hypovolaemia. Other (non-prostaglandin) products of arachidonic acid metabolism, with a similar action (inhibition of Na reabsorption) have also been identified. Their synthesis is catalysed by the cytochrome P-450 enzyme, and like the medullary prostaglandins they limit the extent to which Na reabsorption can be stimulated and so help to prevent hypoxic damage to the renal medullary cells.

Thromboxane A_2 (TXA_2), unlike the other renal prostanoids, is a vasoconstrictor. Little TXA_2 is normally produced, but its synthesis increases following prolonged ureteral obstruction (and thus could reduce the blood flow to an ineffective kidney).

9.7 *Atrial natriuretic peptide and urodilatin*

Cardiac atrial cells (cardiocytes) contain granules in which is stored the precursor (prohormone) of α-human atrial natriuretic peptide (ANP). The atrial genes code for a 'preprohormone' with 152 amino acids. The removal of a hydrophobic sequence from the N-terminal end of this leaves the 126-amino-acid prohormone. The primary released form of the hormone is α-human ANP (28 amino acids), but a number of smaller fragments have been isolated and similar peptides are released from the brain (brain natriuretic peptide, BNP) and from the kidney itself (renal natriuretic peptide or urodilatin).

ANP can be detected in normal plasma and atrial stretch (e.g. as a result of increased effective circulating volume and hence increased venous return) leads to an increased circulating level of ANP.

Many of the effects of ANP are mediated via specific cell-surface receptors which, when ANP binds to them, elicit an increase in the intracellular level of cyclic guanosine monophosphate (cyclic GMP). The major actions of ANP are:

(1) it binds to receptors on the inner medullary collecting duct cells, leading to closure of epithelial sodium channels, and inhibition of $Na^+ K^+$-ATPase, thereby decreasing Na^+ reabsorption at this nephron site;
(2) inhibition of aldosterone secretion by the zona glomerulosa of the adrenal cortex;
(3) reduction of renin release (an effect which will itself reduce aldosterone secretion);
(4) vasodilatation, particularly of the afferent arterioles, leading to increased glomerular filtration rate.

All of these actions contribute to the natriuretic effect of ANP. There is also evidence for a decrease in proximal sodium reabsorption elicited by ANP. This is likely to be due to decreased renin release, which will reduce angiotensin II production, and to increased dopamine release (see below). In addition, ANP inhibits ADH release by the posterior pituitary.

As mentioned above, the kidneys themselves synthesize an ANP analogue (i.e. a peptide with an almost identical amino-acid sequence to that of ANP). This renal natriuretic peptide has been termed **urodilatin**. Wheras ANP has 28 amino acids, urodilatin has an additional four amino acids on its amino terminus end. Its action on medullary collecting duct cells is identical to that of ANP mentioned above – it inhibits the Na^+K^+-ATPase, and so decreases Na reabsorption, and it has been reported that its release as a consequence of increases in effective circulating volume contributes to increased sodium excretion. However, since sodium reabsorption requires metabolic energy, the urodilatin could also have a role resembling that of the medullary prostaglandins mentioned in Section 9.6, i.e. it may limit the extent to which Na reabsorption can be increased and so protect the renal medullary cells against hypoxic damage.

9.8 *Other factors which may be involved in regulating sodium excretion*

9.8.1 *Dopamine*

The kidneys synthesize dopamine, mostly in the cells of the proximal tubule, and this dopamine synthesis reduces tubular sodium transport, by inhibiting the Na^+K^+-ATPase and by decreasing Na^+–H^+ antiport activity. Dopamine is also a vasodilator, but the natriuretic effect is mainly due to its tubular action, since dopamine increases sodium excretion even when renal blood flow and GFR are unaffected.

9.8.2 *Kinins*

Kinins are vasodilator peptides produced from precursor proteins ('kinino-gens') by the enzyme kallikrein. The effects of renally produced kinins are very similar to those of the prostaglandins; they vasodilate the kidney, antagonize ADH-induced increases of urine osmolality and are natriuretic. However, their significance in the overall control of sodium excretion is not clear. The haemo-dynamic effects of kinins are thought to be due to the increased production of nitric oxide (NO).

9.8.3 *Nitric Oxide*

Nitric oxide (NO) is synthesized in the vascular endothelium, and is a vasodilator. In the kidney, nitric oxide synthesis in the afferent arteriolar walls causes dilatation of the arterioles, and so increases glomerular capillary pres-sure and glomerular filtration rate (for mechanism, see Figure 7.4). Increases in dietary sodium intake lead to increased NO synthesis, but this may primarily contribute to blood pressure regulation, and it is currently unclear whether it also plays a part in regulating sodium balance.

9.8.4 *Adrenomedullin*

Adrenomedullin is a 52 amino acid peptide which was originally identified in human pheochromocytoma cells (tumour cells) of the adrenal medulla. The mRNA encoding the adrenomedullin precursor has subsequently been identi-fied in many tissues, including the kidneys, lungs, heart, spleen, duodenum and salivary glands.

In the kidneys, adrenomedullin is present in both glomerular and tubular cells. Administration of adrenomedullin produces a potent natriuresis, via an increase in glomerular filtration rate, and a decrease in tubular sodium reab-sorption, although its role in sodium regulation is still uncertain.

9.8.5 *Endothelin*

Endothelin (ET) is a 21 amino acid peptide which exists in three forms, ET-1, ET-2 and ET-3, which have slightly different amino acid compositions and are separate gene products. ET-1 is the main endothelin synthesized by the kidney. It is a powerful vasoconstrictor and if administered by infusion it profoundly decreases renal blood flow and GFR. Paradoxically, low doses of ET-1 are diuretic and natriuretic, due to an inhibitory effect on proximal tubular sodium reabsorption. ET-1 also inhibits renin release, but whether it has a role in normal sodium homeostasis is still not clear.

9.8.6   *Natriuretic hormone*

Over the past thirty years there have been reports of an as yet unidentified 'natriuretic hormone' – an endogenous ouabain or digoxin-like substance, which is an Na^+K^+-ATPase inhibitor. The evidence for the existence of such a substance remains controversial.

9.9   *ADH and the relationship between osmotic regulation and volume regulation*

In the previous chapter, it was seen that ADH release is controlled by osmoreceptors and that changes in ADH release lead to changes in water reabsorption by the kidney, and thereby regulate body fluid osmolality. However, there are circumstances in which the maintenance of effective circulating volume is more important (to survival) than is the maintenance of body fluid osmolality. So, when the effective circulating volume is threatened, volume regulation takes precedence over osmotic regulation. This is easily demonstrated: the diuresis brought about by water ingestion (i.e. decreased plasma osmolality) is greatly reduced by a modest (non-hypotensive) haemorrhage, i.e., a small reduction in effective circulating volume.

Changes in effective circulating volume are detected by stretch receptors in the right and left atria (afferent fibres are in the vagus nerve) and rapidly lead to appropriate alterations in urine flow (e.g. stretching the left atrial receptors produces a diuresis), mediated by changes in ADH release. In addition to these receptors in the low-pressure parts of the circulation, receptors which can alter ADH release also exist in the arterial (high-pressure) part of the vascular system and it is likely that these are activated by volume reductions which lead to blood pressure decreases.

How do body fluid osmolality and effective circulating volume interact to regulate ADH release? Changes in volume alter the range of plasma osmolalities over which ADH is released, as shown in Figure 9.4. Put another way, decreases in effective circulating volume lead to increased ADH release to retain water and this leads to a decreased plasma osmolality, so that the body accepts a reduced osmolality as the price for maintaining volume at a level higher than it would otherwise be.

It is important to realize that this control over ADH release by volume changes constitutes an 'emergency' mechanism, and that the normal day-to-day control over ADH is via osmoreceptors.

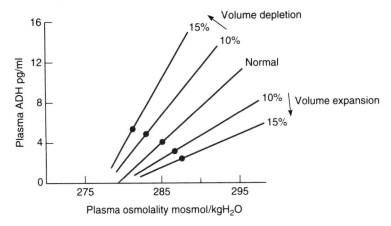

Figure 9.4 The effects of changes in body fluid volume on the regulation of body fluid osmolality by ADH. The black dots on each line show the normal plasma level of ADH for a particular body fluid volume. Thus, decreases in volume cause an adjustment, downwards, in the 'normal' osmolality – the body 'accepts' a lower osmolality (i.e. retains water) in order to minimize the decrease in volume.

9.10 *Overall scheme of body fluid volume regulation*

The obvious point which emerges from the foregoing discussion is that no *single* factor can be said to regulate sodium ion excretion. Instead, many factors and mechanisms are involved and this provides a 'failsafe' system, whereby abnormalities or malfunctions affecting one factor are unlikely to disturb overall sodium balance.

For example, kidney transplantation is a widely performed operation and transplanted kidneys are denervated. Nevertheless, they are able to maintain sodium balance and adjust sodium excretion according to the needs of the body. The fact that the renal nerves are not essential does not mean they have no function. If kidneys are denervated *in situ*, then a diuresis occurs (denervation diuresis) which usually lasts for some hours. This suggests that the nerves are normally playing a part in keeping urine flow (and sodium excretion) low, but that when the renal nerves are sectioned other control mechanisms can, after a time, compensate for the renal nerves.

Similar reasoning can be applied in the case of aldosterone. A certain minimum amount of aldosterone is necessary for life, but the absolute level is not critical for a steady state of sodium balance to be achieved.

Figure 9.5 shows the main factors involved in the regulation of body fluid volume.

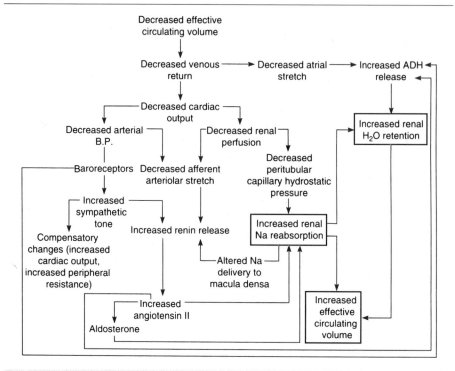

Figure 9.5 The regulation of body fluid. It should be noted that, in the scheme illustrated (the effects of decreased effective circulating volume), no role is shown for atrial natriuretic peptide. This is because the normal plasma level of ANP is low and decreases below the normal value are unlikely to have any effect. ANP may be important (see text) if the body fluid volume (and effective circulating volume) increase. Also not shown in the diagram is the possibility that the GFR may change if volume disturbances are sufficiently large (i.e. GFR decreasing if effective circulating volume decreases).

Further reading

Brenner, B. M., Ballermann, B. J., Gunning, M. E. and Zeidel, M. L. (1990) Diverse biological actions of atrial natriuretic peptides. *Physiol. Rev.* **70**, 665–699

Brooks, D. P. (1997) Endothelin: The prime suspect in kidney disease. *News Physiol. Sci.* **12**, 83–89

Carretero, O. A. and Scicli, A. G. (1988) Kinins paracrine hormones. *Kidney Int.* **34** (Suppl. 26) 26–59

Clausen, T. (1998) Clinical and therapeutic significance of the Na^+, K^+ pump. *Clin. Sci.* **95**, 3–17

DiBona, G. F. (1989) Neural control of renal function. Cardiovascular implications. *Hypertension* **13**, 539–548

Synthesis of renin

Renin is a neutral protease enzyme, molecular weight 42000, and the pH optimum for the renin-catalysed cleavage of angiotensinogen is 7.0. The gene for renin has been identified and the genomic sequence elucidated. Renin is initially synthesized as a large precursor, pre-pro-renin, with 340 amino acids. This is cleaved by the removal of 23 amino acids from the N-terminal end to form prorenin, which is then glycosylated.

In the juxtaglomerular cells, prorenin is synthesized continually and most of this (inactive) is secreted. A proportion remains in the cells and is converted into renin, which is stored in the secretory granules and released as required, by the stimuli mentioned in the text.

Active renin is secreted only by the juxtaglomerular cells; prorenin is also synthesized by other organs (including adrenal glands, ovary, testis, pituitary), but there is no evidence that prorenin can be converted into renin in the circulation. Thus the granular cells of the juxtaglomerular apparatus appear to be the only source of active renin.

Gesek, F. A. and Friedman, P. A. (1995) Sodium entry mechanisms in distal convoluted tubule cells. *Am. J. Physiol.* **268**, F89–F98

Kone, B. C. and Baylis, C. (1997) Biosynthesis and homeostatic roles of nitric oxide in the normal kidney. *Am. J. Physiol.* **272**, F561–F578

Knox, F. G., Burnett, J. C., Kohan, D. E., Spielman, W. S. and Strand, J. C. (1980) Escape from the sodium retaining effects of mineralocorticoids. *Kidney Int.* **17**, 263–276

Lote, C. J. and Haylor, J. (1989) Eicosanoids in renal function. *Prostaglandins, Leukotrienes and Essential Fatty Acids* **36**, 203–217

Matsusaka, T. and Ichikawa, I. (1997) Biological functions of angiotensin and its receptors. *Annu. Rev. Phsiol.* **59**, 395–412

Meyer, M., Richter, R., Brunkhorst, R., Wrenger, E., Schulz-Knappe, P., Kist, A., Mentz, P., Brabant, E. G. and Koch, K. M. (1996) Urodilatin is involved in sodium homeostasis and exerts sodium state-dependent natriuretic and diuretic effects. *Am. J. Physiol.* **271**, F489–F497

Wilkins, M. R., Nunez, D. J. and Wharton, J. (1993) The natriuretic peptide family: turning hormones into drugs. J. *Endocrinol.* **137**, 347–359

Renal regulation of body fluid pH

10

10.1 *Introduction*

In Chapter 1 it was pointed out that normal metabolism requires the extra- and intracellular fluid compositions to be kept relatively constant; this constancy includes the hydrogen ion concentration, conventionally measured as pH. The pH must be maintained within fairly narrow limits: the normal blood pH is 7.40 and in health is kept within the range 7.35–7.45, although a range of 7.0–7.8 can be tolerated. If the pH falls as low as 6.8, recovery is almost impossible.

It is not entirely clear why a pH of 7.4 is so important. Most of the enzyme reactions in the body have a pH optimum, but the pH dependence of such reactions is much less sharply defined than the pH dependence of the whole organism. However, by using the pH notation, we tend to obscure the range of variation of H^+ concentration which can be tolerated.* Thus the range of pH from 7.35 to 7.45 is an $[H^+]$ range of 45–35 nmol/l, which is a change of over 20% (it should be borne in mind that a 20% change in, for example, body sodium concentration would have very serious consequences). Nevertheless, the hydrogen concentration in the body is tiny. The normal sodium concentration of the plasma (140 mmol/l) is three and a half *million* times the normal hydrogen concentration (40 nmol/l).

10.2 *Physiological buffers*

A **buffer solution** is one which, when acid or base is added to it, minimizes the change of pH. Buffer solutions consist of a weak acid and the conjugate

---

*pH $= -\log [H^+]$, or, to put it another way, pH is the logarithm of the reciprocal of the H^+ concentration.

base of that acid, i.e.

$$HA \rightleftharpoons H^+ + A^-$$
acid　　proton　conjugate base

or, alternatively, the buffer solution may be a weak base and its conjugate acid:

$$B + H^+ \rightleftharpoons BH^+$$
base proton conjugate acid

From these equations it is apparent that acids can donate protons, whereas bases can accept them. Several physiological buffers can be both acids and bases – they can accept or donate protons – and such substances are known as amphoteric. Amino acids and phosphate can behave in this way.

10.2.1 *pK values and equilibrium constants*

The equilibrium for any reaction can be defined by an equilibrium constant, K. Thus for the reaction

$$HA \rightleftharpoons A^- + H^+$$

$$K = \frac{[H^+][A^-]}{[HA]}$$

or in general terms

$$K = \frac{[H^+][base]}{[acid]} \tag{1}$$

However, K is generally a very small figure and it is therefore more convenient to use the term pK. This is analogous to pH, i.e. it is the log of $1/K$.

$$pK = \log\frac{1}{K} = -\log K$$

$$pH = \log\frac{1}{[H^+]} = -\log [H^+]$$

If we rearrange Equation (1)

$$H^+ = \frac{K[acid]}{[base]}$$

we obtain what is known as the Henderson equation. Changing this into the pH and pK notation, we can derive

$$pH = pK + \log\frac{[base]}{[acid]}$$

which is known as the Henderson–Hasselbalch equation.

Table 10.1 The main physiological buffer systems

Body compartment	Buffer
Blood	Bicarbonate/CO_2
	Haemoglobin (HHb/Hb^- and $HHbO_2/HbO_2^-$)
	Plasma proteins (H^+ Protein/Protein$^-$)
	Phosphate ($H_2PO_4^-$, HPO_4^{2-})
Extracellular fluid	Bicarbonate/CO_2
and cerebrospinal	Proteins (H^+ Protein/Protein$^-$)
fluid	Phosphate ($H_2PO_4^-/HPO_4^{2-}$)
Intracellular fluid	Proteins (H^+ Protein/ Protein$^-$)
	Phosphate ($H_2PO_4^-/HPO_4^{2-}$)
	Organic phosphates
	Bicarbonate/CO_2

Source: Modified, with permission, from Gardner, M. L. G., Medical *acid–base balance: the basic principles* (Baillière Tindall, London, 1978).

10.2.2 *Physiological buffering*

In the body fluids, there are several buffer systems (Table 10.1). In blood, the main ones are the bicarbonate system, haemoglobin and the plasma proteins. Within the cells, the buffers include bicarbonate, phosphate and proteins. Throughout the body fluids, the bicarbonate buffer system is of primary importance. The reaction sequence for this system is:

$$CO_2 + H_2O \rightleftharpoons H_2CO_3 \rightleftharpoons H^+ + HCO_3^-$$

The great importance of this system stems from the precise regulation of the CO_2 concentration by the lungs and the bicarbonate concentration by the kidneys. For the bicarbonate buffer system, we can write the Henderson–Hasselbalch equation as follows:

$$pH = pK + \log \frac{[HCO_3^-]}{[H_2CO_3]}$$

but since the $[H_2CO_3]$ is proportional to the pCO_2 (partial pressure of CO_2 in mmHg), this can be written as

$$pH = pK + \log \frac{[HCO_3^-]}{0.03 \times pCO_2}$$

Substituting the normal values of HCO_3^- concentration (24–25 mM) and pCO_2 (40 mmHg), and with the pK of the reaction,

$$H_2CO_3 \rightarrow H^+ + HCO_3^-$$

being 6.1, this becomes

$$pH = 6.1 + \log \frac{24}{0.03 \times 40}$$

i.e.

$$= 6.1 + 1.3$$
$$= 7.4$$

To simplify the following discussion still further, we could state that

$$pH \propto \frac{HCO_3^-}{pCO_2}$$

and since bicarbonate is regulated by the kidneys and pCO_2 by the lungs, it is apparent that pH depends on the activity of both organs.

One further important point needs to be made before we consider how the renal regulation occurs. This point is that, by precisely fixing the [base]/[acid] ratio for one buffer system (the bicarbonate system), the pH of the body fluids is determined. This pH will then determine the [base]/[acid] ratio for all the other buffer systems in the body. The bicarbonate system, chemically, is a poor buffer, but physiologically it is extremely effective because of the control over $[HCO_3^-]$ and pCO_2.

10.3 *Renal regulation of plasma bicarbonate concentration*

The metabolism of our food generates H^+ which is largely removed from the body by ventilation, since the H^+ reacts with HCO_3^- to produce CO_2 which is exhaled. In this process, bicarbonate (HCO_3^-) is lost from the body as CO_2 and the role of the kidney can thus be seen as the conservation of the remaining HCO_3^- and the generation of additional HCO_3^-.

Table 10.2 shows the normal pCO_2, pH, $[H^+]$ and $[HCO_3^-]$ in the blood. It can be seen that bicarbonate is normally present in the plasma at a concentration of approximately 25 mmol/l. Bicarbonate ion, HCO_3^- is freely filtered at the glomeruli, hence the concentration of bicarbonate entering the nephron is also 25 mmol/l. The kidney behaves as if there is a T_m for bicarbonate reabsorption (as shown in Figure 10.1) with the T_m^* set very close to the amount which is filtered at the normal plasma concentration. This provides a means of dealing with increases in plasma $[HCO_3^-]$, since an increase will lead to the T_m being exceeded and HCO_3^- being excreted until the plasma level is again too low to exceed the T_m.

*In fact the T_m for HCO_3^- is variable. It can be adjusted by the rate of H^+ secretion (see Chapter 5), but it also varies directly with the fractional sodium reabsorption.

Table 10.2 Normal values for the bicarbonate buffer system in blood

	[H$^+$] (nmol/l)	pH	pCO$_2$ (mmHg)	[HCO$_3^-$] (mmol/l)
Arterial	40	7.4	40	24
Venous	46	7.35	46	25

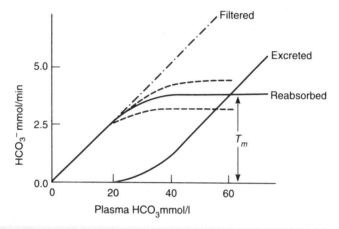

Figure 10.1 The renal handling of HCO$_3^-$. There is an apparent T_m for HCO$_3^-$ reabsorption, but this T_m is variable, as it depends primarily on the rate of tubular H$^+$ secretion. The limits of the variability are shown by the dashed lines. There is a similar degree of variability in the amount of HCO$_3^-$ excreted at any particular filtered load, but for clarity this is not shown in the diagram.

In the proximal tubule, most (90%) of the filtered bicarbonate is absorbed from the tubular fluid. This bicarbonate reabsorption is in fact brought about not by the transport of HCO$_3^-$ ions, but by the luminal conversion of HCO$_3^-$ into CO$_2$ (Figure 10.2). Hydrogen (H$^+$) is secreted from the proximal tubule cell into the lumen, where it associates with HCO$_3^-$ to form H$_2$CO$_3$. This carbonic acid dissociates into CO$_2$ and H$_2$O. The reaction is catalysed by an enzyme, **carbonic anhydrase**, which is present in the brush borders of the cells, and so catalyses the luminal reaction without being lost in the urine.

The CO$_2$ so formed can readily diffuse into the tubule cells, where intracellular carbonic anhydrase catalyses the rehydration of CO$_2$ to H$_2$CO$_3$, which dissociates to H$^+$ (available for secretion) and HCO$_3^-$ which re-enters the plasma via the peritubular fluid. We might therefore, claim that the purpose of H$^+$ secretion is to permit HCO$_3^-$ reabsorption. Under normal circumstances, less than 0.1% of filtered HCO$_3^-$ is excreted in the final urine.

Although the entry of Na$^+$ into the proximal tubule cells is down the electrochemical gradient, Na$^+$ entry is coupled to H$^+$ secretion (i.e. there is an antiport or countertransport process) and the transport is electroneutral. In

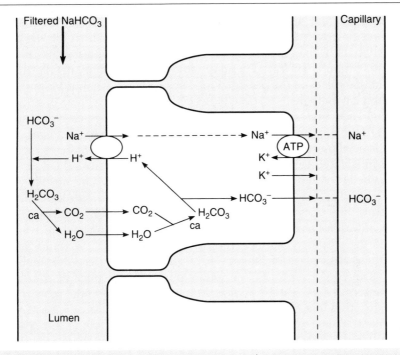

Figure 10.2 Proximal tubular HCO_3^- reabsorption. H^+, secreted into the tubular lumen (left side of diagram) by antiport (with Na^+) and by the proton ATPase, combines with filtered HCO_3^- to form carbonic acid (H_2CO_3) which is in turn rapidly converted into CO_2 and H_2O, catalysed by carbonic anhydrase (present in the brush border of the cells). CO_2 which enters the cells can be rehydrated to carbonic acid and hence can form HCO_3^-, some of which enters the blood (right side of diagram). Thus CO_2 entry into the cells in this way is, in effect, HCO_3^- reabsorption. However, some of the HCO_3^- generated from CO_2 in the cells is secreted into the lumen in exchange for Cl^- (not shown, see text). In the distal nephron, secreted H^+ can combine with luminal HCO_3^-, but there is no luminal carbonic anhydrase, hence there is little conversion of H_2CO_3 to CO_2 and H_2O, and so little bicarbonate reabsorption. Secreted H^+ also combines with anions other than HCO_3^- (see Figures 10.3 and 10.4). ca, carbonic anhydrase.

addition, the luminal membrane has a primary H^+ secretory mechanism, using proton ATPase, although this is of little importance in the proximal tubule.

However, there is evidence that most apical membrane Na^+ reabsorption is associated with H^+ secretion, which generates CO_2 in the tubular lumen. The CO_2 enters the cells and again forms HCO_3^- (Figure 10.2), much of which then exchanges with luminal Cl^-, i.e. there is HCO_3^- secretion and Cl^- reabsorption, such that the overall effect is predominantly the reabsorption of NaCl (see also Figure 5.1 and p. 54).

The exit of HCO_3^- from the cells across the basal and basolateral membranes is via a Na^+–HCO_3^- cotransporter which transports out three HCO_3^- ions for each Na^+ ion.

The fundamental acid–base regulating process that occurs in the 'distal tubule' and collecting duct is the same as that in the proximal tubule, viz. it is H^+ secretion. However, the process differs in a number of respects from that in the proximal tubule. The distal nephron processes are, quantitatively, much less important than those in the proximal tubule, since they account for only 10% of the total bicarbonate reabsorption. However, in order to achieve this, the distal tubule and collecting duct have to secrete H^+ against a much bigger gradient than that in the proximal tubule and this requires specific mechanisms.

The distal nephron cell types which are involved in acid–base regulation, are the **intercalated cells** (see p. 26) which secrete H^+ across the apical membrane by the H^+-ATPase and by an H^+K^+-ATPase (which pumps H^+ out and K^+ in). The linked Na^+/H^+ transport of H^+ is of lesser importance in the distal nephron, but does occur in the principal cells and is one of the means whereby aldosterone enhances H^+ secretion (see Figure 6.8).

In both the proximal tubule and the distal nephron, H^+ secretion leads not only to the reabsorption of filtered HCO_3^-, it also generates additional HCO_3^- for the plasma. In order for H^+ secretion to occur, there must be some way in which the secreted H^+ can be 'mopped up' (i.e. buffered) in the tubular lumen, in order to provide a continuing gradient for secretion, or at least to prevent the gradient against which secretion is occurring from being too large. It is mainly filtered HCO_3^- which does this H^+ buffering, but there are also other forms of combination for the secreted H^+.

10.3.1 *Conversion of alkaline phosphate to acid phosphate*

In the plasma there are two phosphate salts, disodium hydrogen phosphate (alkaline phosphate, Na_2HPO_4) and sodium dihydrogen phosphate (acid phosphate, NaH_2PO_4). The ratio of alkaline phosphate to acid phosphate in plasma is approximately $4:1$. This ratio is determined by the plasma pH according to the Henderson–Hasselbalch equation.

$$H^+ + HPO_4^{2-} \rightleftharpoons H_2PO_4^-$$
$$\text{proton} \quad \text{base} \qquad\qquad \text{acid}$$

The pK for this reaction is 6.8, so

$$pH = 6.8 + \log \frac{[HPO_4^{2-}]}{[H_2PO_4^-]}$$

and where $pH = 7.4$, the ratio $[HPO_4^{2-}]/[H_2PO_4^-]$ is $4:1$.

Both the acidic and basic forms of phosphate are filtered at the glomerulus and, because of H^+ secretion into the nephron, the ratio $[HPO_4^{2-}]/[H_2PO_4^-]$ is reduced, i.e. alkaline phosphate is converted into acid phosphate. This will occur to some extent in the proximal tubule (since there is a small fall in pH of the tubular fluid proximally), but the main conversion will occur in the distal tubule. The process is shown diagrammatically in Figure 10.3. The $H_2PO_4^-$

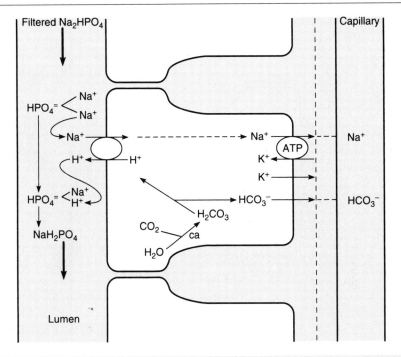

Figure 10.3 Conversion of alkaline phosphate to acid phosphate in the tubular lumen (left side of diagram). Secreted H^+ in both the proximal and distal tubules combines with filtered alkaline phosphate (Na_2HPO_4) to convert it into acid phosphate (NaH_2PO_4) as Na^+ is reabsorbed. In the proximal tubule, H^+ secretion is by antiport with Na^+ and by the proton ATPase, whereas in the distal nephron only the proton ATPase secretory mechanism is important. The intracellular production of H^+ for secretion generates an HCO_3^- ion which enters the plasma.

constitutes the **titratable acidity** of the urine. The important point about this is that the secretion of H^+ generates HCO_3^- for the plasma.

10.3.2 *Ammonia secretion*

In most parts of the nephron, but mainly in the proximal tubule cells, ammonium ions, NH_4^+, are produced as a result of the deamination of glutamine to glutamic acid and to 2-oxoglutarate (α-ketoglutarate):

$$glutamine \rightarrow 2\text{-}oxoglutarate + 2NH_4^+$$

The NH_4^+ is secreted into the tubular lumen (Figure 10.4). The 2-oxoglutarate left behind in the tubule cells reacts with H^+ (generated from $CO_2 + H_2O$) to form glucose or CO_2, leaving bicarbonate, HCO_3^- to enter the plasma. Thus the deamination of glutamine, generating NH_4^+, produces bicarbonate for the plasma. If the NH_4^+ were not secreted, but remained in the body,

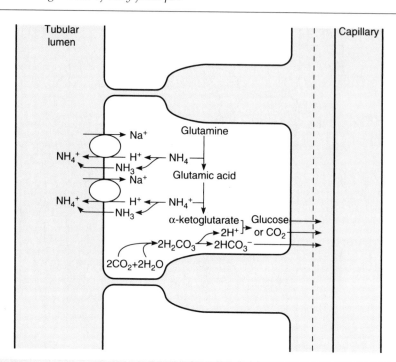

Figure 10.4 NH_4^+ production and secretion by tubular cells and consequent production of HCO_3^- for the plasma.

then the liver would utilize NH_4^+ and HCO_3^- to produce urea for excretion (with no gain of HCO_3^- for the plasma). Thus the greater the renal secretion of NH_4^+, the greater the HCO_3^- production for the plasma.

A large fraction (~50%) of the NH_4^+ secreted into the lumen by the proximal tubule cells, is reabsorbed by the thick ascending limb of the loop of Henle and accumulates in the medullary interstitium This absorption is by $Na^+/NH_4^+/2Cl^-$ cotransport across the apical membrane (i.e. NH_4^+ substitutes for K^+ on the transporter). The NH_4^+ absorption lowers the lumenal NH_4^+ concentration and hence also lowers the NH_3 concentration as the removal of NH_4^+ shifts the $NH_4^+ \rightleftharpoons NH_3 + H^+$ buffer reaction towards NH_4^+.

The NH_4^+ from the medullary interstitium is then secreted as NH_3 by the cells of the collecting tubules (non-ionic diffusion) and converted into NH_4^+ in the collecting tubule lumen by combination with secreted H^+ (Figure 10.5). NH_4^+ excretion is increased in acidosis. This is because:

(a)   the enzymes in the proximal tubule which deaminate glutamine are stimulated by acidosis;

(b) NH_3 conversion to NH_4^+ in the collecting tubule is greater if H^+ secretion is greater, and this maintains a gradient for NH_3 secretion so that more NH_3/NH_4^+ is removed from the renal medulla.

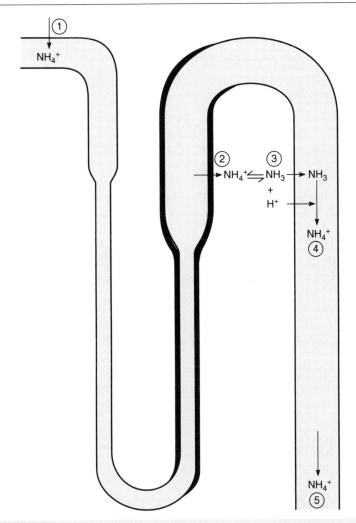

Figure 10.5 Overall scheme of renal NH_3/NH_4^+ handling. ① NH_4^+ production in proximal tubule (see Figure 10.4). ② Much of the NH_4^+ is reabsorbed by the thick ascending limb of the loop of Henle. This reabsorption occurs because NH_4^+ substitutes for K^+ in the $Na^+ : K^+ : 2Cl^-$ cotransporter in the apical cell membrane (see Figure 6.2). ③ In the medullary interstitium there is an equilibrium between NH_4^+ and NH_3. As NH_3 is a small uncharged molecule, it can readily cross cell membranes and can enter the collecting tubule cells and be secreted into the tubular lumen. ④ Since the cells also secrete H^+ (by the H^+-ATPase), NH_3 is converted into NH_4^+ in the tubular lumen. This conversion maintains a gradient for NH_3 secretion, and since NH_4^+ cannot readily diffuse back across the cell membranes, it is trapped in the tubular lumen and excreted ⑤ in the urine.

10.3.3 *Summary*

H^+ secretion in the nephron leads to:

(a)   bicarbonate reabsorption into the plasma;
(b)   the generation of further bicarbonate to enter the plasma.

H^+ secretion demands that H^+ must be able to combine with anions in the tubule. For the process of bicarbonate reabsorption, H^+ combines with bicarbonate itself. For the process of bicarbonate generation, H^+ combines with HPO_4^{2-} or NH_3.

Because of H^+ secretion, the pH of the tubular fluid falls progressively along the nephron. This fall is small in the proximal tubule (from 7.40 down to about 6.90), but the pH may be as low as 4.5 in the collecting duct.

Although the urine is usually acidic and can be titrated to determine the 'titratable acidity', this only constitutes a fraction of the total H^+ secretion, because: total H^+ secretion $\equiv HCO_3^-$ reabsorption $+ H_2PO_4^-$ excretion $+ NH_4^+$ excretion; and only the $H_2PO_4^-$ excretion is 'titratable acidity'.

10.4   *Regulation of H^+ secretion according to acid–base balance requirements*

Acid–base disturbances can be divided into two categories, each with two sub-categories; (1) disturbances of respiratory origin: (i) respiratory acidosis, and (ii) respiratory alkalosis; (2) disturbances of non-respiratory origin: (i) metabolic acidosis, and (ii) metabolic alkalosis. (The term 'metabolic' refers to acid–base disturbances which affect the bicarbonate carbonic acid buffer system by a means other than an alteration of pCO_2. Such disturbances are frequently due to diet rather than metabolism *per se*.)

In each of the four disturbances, there is initially a change in body fluid pH (i.e. a change in H^+ concentration). However, the buffer and compensatory systems are so effective that a change in pH may be barely measurable.

Changes in body fluid pH will include changes in arterial pH and, when changes in extracellular fluid pH occur, there are parallel (although not necessarily identical) changes in intracellular pH. Therefore a change in arterial pH is reflected in the pH of all the cells of the body, including the renal tubule cells. The rate of H^+ secretion from the tubule cells varies inversely with pH (i.e. varies directly with H^+ concentration). This is vital for the compensation for, and correction of, acid–base disturbances.

10.4.1 *Compensation versus correction in acid–base disturbances*

When an acid–base disturbance occurs, compensatory mechanisms immediately come into play to minimize and correct the pH change. However, these compensatory mechanisms do not restore acid–base balance to normal; they only restore pH to normal.

Normal acid–base status is not only a pH of 7.4, but is also a plasma $[HCO_3^-]$ of about 25 mmol/l and a plasma pCO_2 of about 40 mmHg. An acid–base disturbance disturbs at least two of these three variables. **Compensation** is the restoration of pH towards normal even though $[HCO_3^-]$ and/or pCO_2 are still disturbed. **Correction** is the restoration of normal pH, $[HCO_3^-]$ and pCO_2. In fact, we might say that the purpose of the body's acid-balance regulating mechanisms is the maintenance of normal pH. The tools for this are pCO_2 and HCO_3^-. The regulation of pCO_2 and $[HCO_3^-]$ *per se*, although of importance, is a subordinate function, which is sacrificed when necessary in the interests of pH regulation.

In the following pages, much use will be made of the graph of plasma HCO_3^- concentration versus plasma pH. On this graph, pCO_2 isobars are shown (i.e. the partial pressure of CO_2 in mmHg), as in Figure 10.6. Because of the relationship between pH, HCO_3^- concentration and pCO_2 given by the equation

$$pH \propto \frac{[HCO_3^-]}{pCO_2}$$

if we know the value of any two of the variables, then there is only one possible value for the third variable, e.g. if we know pCO_2 and pH, there can be only one possible HCO_3^- concentration. The graph depicts this situation diagrammatically.

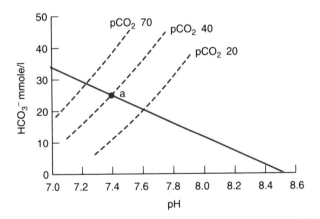

Figure 10.6 Graph of plasma HCO_3^- concentration versus plasma pH for normal whole blood (i.e. plasma + red cells). pH, $[HCO_3^-]$ and pCO_2 are interrelated and a knowledge of any two of the values enables us to determine the third. Point (a) is the normal point, i.e. pH 7.4 (normal range 7.3–7.45), plasma $[HCO_3^-]$ 25 mM (normal range 2–27), pCO_2 40 mmHg (normal range 35–48 mmHg). In the following diagrams (Figures 10.7–10.10), point (a) is the normal point, point (b) is the uncompensated acid–base disturbance and point (c) is an acid–base disturbance with compensation.

Let us now examine the types of acid–base disturbance in detail. The distinction between 'compensation' and 'correction' should become clearer in the following sections.

In these sections, the simplified Henderson–Hasselbalch equation is used to show changes in pH, HCO_3^- and pCO_2, and three types of arrow are used to indicate changes: ↑ or ↓ is the initial acid–base defect; ↑ or ↓ is the consequence of the defect; ⬆ or ⬇ is compensation.

10.4.2 *Respiratory acidosis*

This is a disturbance of acid–base balance which occurs when the respiratory system is unable to remove sufficient CO_2 from the body to maintain normal pCO_2. So the reaction is displaced to the right by the high pCO_2 (law of mass action):

$$CO_2 + H_2O \rightleftharpoons H_2CO_3 \rightleftharpoons H^+ + HCO_3^-$$

The consequence of this defect is an increased $[H^+]$ (i.e. acidosis-reduced pH), and an increased $[HCO_3^-]$. These changes are expressed graphically in Figure 10.7.

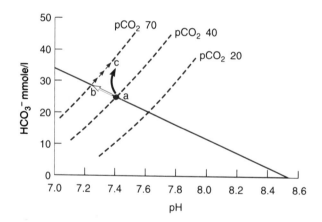

Figure 10.7 Respiratory acidosis and its compensation. The defect in respiratory acidosis is an increase in the blood pCO_2, which increases the plasma $[HCO_3^-]$ and lowers the pH. These changes move the values of the three variables (pCO_2, pH, $[HCO_3^-]$) from the normal point (a) to point (b). Removal of the defect (i.e. a lowering of the pCO_2) would return $[HCO_3^-]$ and pH to normal. However, if the defect persists, then there is compensation, which serves to restore the pH towards normal. The compensation process is the renal retention of HCO_3^-, so that, on the graph, the values change from those of point (b), to those of point (c). Thus the defect (increased pCO_2) raises $[HCO_3^-]$, but compensation raises $[HCO_3^-]$ still further. This illustrates the important point that it is pH which is regulated. $[HCO_3^-]$ and pCO_2 are the 'tools' for the regulation of pH. (Generally, compensation occurs as the defect develops, so that the true change in the variables may occur along the line shown by the direct a–c arrow.)

In the simplified form of the Henderson–Hasselbalch equation, we have:

$$\downarrow pH \propto \frac{[HCO_3^-]\uparrow}{pCO_2 \uparrow}$$

A change in [H$^+$] in the body fluids changes the rate of H$^+$ secretion from renal tubular cells. In respiratory acidosis, [H$^+$] is increased, therefore the rate of H$^+$ secretion also increases. This increased secretion is sufficient to reabsorb the filtered HCO$_3^-$ (even though the plasma [HCO$_3^-$] is raised by the defect and therefore the amount of HCO$_3^-$ filtered is increased) and to generate further HCO$_3^-$ for the plasma. This increased renal H$^+$ secretion leading to increased plasma [HCO$_3^-$] is **renal compensation for respiratory acidosis**. It is shown graphically in Figure 10.7, and in the Henderson–Hasselbalch (simplified) equation below:

$$\uparrow\downarrow pH \propto \frac{[HCO_3^-]\uparrow\uparrow}{pCO_2 \uparrow}$$

It should be noted that compensation restores the pH towards normal, but plasma [HCO$_3^-$] is raised (both as a consequence of the defect, and by compensation) and pCO_2 is raised (the initial defect). To restore normal [HCO$_3^-$] and pCO_2 would require a respiratory alteration, to lower pCO_2 (i.e. correction of the defect).

Respiratory acidosis (hypercapnia) is a pCO_2 of greater than about 45 mmHg in arterial blood.

Causes of respiratory acidosis
The commonest causes of respiratory acidosis are chronic bronchitis and emphysema. Obstructions of the airway, for example by a foreign body, tumour or a constriction (in bronchial asthma), will also impair lung gas exchange and so raise arterial pCO_2, and mechanical injuries of the chest may impair respiration. Respiratory acidosis may also occur as a result of injuries and infections directly affecting the respiratory centre in the brain stem. General anaesthetics, morphine and barbiturates are respiratory-centre depressants.

Other possible causes of respiratory acidosis include defective pulmonary diffusion (but this has a more marked effect on pO_2 than on pCO_2) and inadequate or non-uniform lung perfusion (although this too usually affects primarily the pO_2, and so may cause respiratory alkalosis, see below).

10.4.3 *Respiratory alkalosis*

This is caused by the excessive removal of CO_2 from the body by the respiratory system, so that arterial pCO_2 falls below about 35 mmHg. The reaction:

$$CO_2 + H_2O \rightleftharpoons H_2CO_3 \rightleftharpoons H^+ + HCO_3^-$$

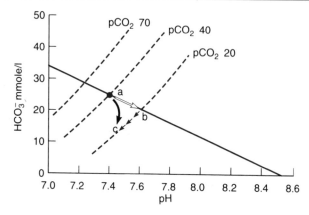

Figure 10.8 Respiratory alkalosis and its compensation. The defect in respiratory alkalosis is decreased arterial pCO_2, which lowers the plasma [HCO$_3^-$] and raises pH. Thus the values are shifted from point (a) to point (b). The increased pH reduces the rate of renal H$^+$ secretion, so that less bicarbonate can be reabsorbed or regenerated by the kidney. This leads to compensation, i.e. the plasma [HCO$_3^-$] falls, thereby lowering the pH towards normal (point c). If the defect develops slowly, the compensation will occur progressively as shown by the direct a–c arrow.

is displaced to the left by the lowering of pCO_2 and this leads to a decrease in [H$^+$] (i.e. a pH increase-alkalosis), and a decrease in [HCO$_3^-$]. The changes are shown graphically in Figure 10.8 and in the simplified Henderson–Hasselbalch equation we have

$$\uparrow pH \propto \frac{[HCO_3^-]\downarrow}{pCO_2\downarrow}$$

The decreased pCO_2 and consequent decreased [H$^+$] in the renal tubule cells reduces the rate of H$^+$ secretion, so that, although the plasma [HCO$_3^-$] is reduced, thereby reducing the amount of HCO$_3^-$ filtered by the kidney, the rate of H$^+$ secretion is insufficient to reabsorb all the filtered bicarbonate, or to generate more bicarbonate. Thus HCO$_3^-$ is excreted in the urine, and the plasma [HCO$_3^-$] falls further.

The reduced H$^+$ secretion in the renal tubules, leading to HCO$_3^-$ excretion, is **renal compensation for respiratory alkalosis** and is shown graphically in Figure 10.8. In the equation, we have:

$$\downarrow\uparrow pH \propto \frac{[HCO_3^-]\downarrow\downarrow}{pCO_2\downarrow}$$

The renal compensation restores pH to (or towards) normal, but it lowers the plasma [HCO$_3^-$] (which was lowered anyway as a consequence of the defect) still further. pCO_2 is still reduced (the defect). The restoration of normal plasma [HCO$_3^-$] and pCO_2 requires the removal of the respiratory defect, i.e. requires a reduction of ventilation.

Causes of respiratory alkalosis

Respiration serves to control the arterial pO_2 and pCO_2, and both oxygen and carbon dioxide participate in the control of respiration. It is not appropriate in a renal physiology book to go into the control of respiration in detail; however, in order to understand the usual cause of respiratory alkalosis, some knowledge of respiratory control is essential.

Normally, the maintenance of an adequate pCO_2 occurs automatically if respiration is adequate for the maintenance of normal pCO_2. Increases in arterial pCO_2 increase H^+ concentration in most cells of the body, and this provides the stimulus (at brain-stem chemoreceptors) for an increased ventilation. Decreases in arterial pCO_2 decrease ventilation.

However, if the pO_2 in the inspired air is below normal, so that arterial pO_2 falls significantly, then oxygen lack, acting mainly via chemoreceptors in the carotid body, increases ventilation. When this occurs, the arterial pCO_2 falls. This is respiratory alkalosis.

So hypoxia, leading to increased respiration, can lead to hypocapnia and respiratory alkalosis. This occurs in normal people when they ascend to a high altitude (10 000 feet or more). Other causes of respiratory alkalosis include hyperventilation, which can be due to fever or brain-stem damage (affecting the pons) or to hysterical over-breathing.

10.4.4 *Metabolic acidosis*

This is acidosis which is not caused by a change in the arterial pCO_2. So, in the reaction sequence

$$CO_2 + H_2O \rightleftharpoons H_2CO_3 \rightleftharpoons H^+ + HCO_3^-$$

we can regard the metabolic acidosis as the addition of H^+ to the right of the reaction, so driving it to the left and depleting the plasma HCO_3^- as it does so. (The direct loss of HCO_3^- will also cause metabolic acidosis, as will be clear from the Henderson–Hasselbalch equation below.) Since respiration is unimpaired, we can consider that this sequence of events takes place initially at a constant (normal) pCO_2.

The acidosis is shown in Figure 10.9. In the equation, we have

$$\downarrow pH \propto \frac{[HCO_3^-]\downarrow}{pCO_2}$$

The change in pH, acting on the (peripheral) chemoreceptors, stimulates respiration so that arterial pCO_2 falls. This is **respiratory compensation for metabolic acidosis**. However, the lowering of pCO_2 moves the reaction

$$CO_2 + H_2O \rightleftharpoons H_2CO_3 \rightleftharpoons H^+ + HCO_3^-$$

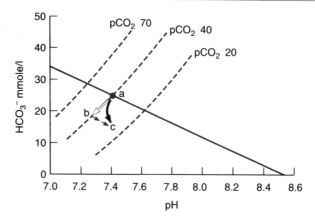

Figure 10.9 Metabolic acidosis and its compensation. Metabolic acidosis, the presence of excess H^+ in the body (from any source other than CO_2), can be regarded as occurring initially at a constant pCO_2. So, in the development of metabolic acidosis, we move from point (a) to point (b) along the $pCO_2 = 40$ isobar. As explained in the text, plasma [HCO_3^-] falls. The correction of the defect, to return from point (b) back to point (a), occurs as a result of the renal excretion of the excess H^+. However, as it is obviously impossible for all of the excess H^+ to be excreted instantaneously, there is respiratory compensation for the metabolic acidosis. Acidosis stimulates the chemoreceptors, which control respiration, and brings about increased ventilation. This lowers the pCO_2 and also lowers plasma [HCO_3^-], so we move from point (b) to a new point, (c), by moving parallel to the normal HCO_3^-/pH line, in the direction of a lower pCO_2 isobar. Generally, the onset of a metabolic acidosis is immediately accompanied by respiratory compensation, so that the values of [HCO_3^-], pH and pCO_2 change as shown by the direct a–c arrow.

to the left thus lowering [H^+] and so raising the pH towards normal, but also further lowering the plasma [HCO_3^-] (Figure 10.9), i.e.

$$\uparrow\downarrow pH \propto \frac{[HCO_3^-]\downarrow\downarrow}{pCO_2\downarrow\downarrow}$$

The removal of the defect is brought about because of the change in pH caused by the defect itself: the increased [H^+] in the blood is reflected in an increased [H^+] in the renal tubular cells, so that the rate of H^+ secretion is increased. This permits the reabsorption of all the filtered HCO_3^- (the filtered HCO_3^- load is below normal), and the regeneration of more HCO_3^- so that the depleted plasma HCO_3^- is restored, thereby restoring the pH to normal, so that respiration can also be normalized.

Causes of metabolic acidosis
Metabolic acidosis can be caused by the ingestion of acids (H^+), or the excessive metabolic production of H^+, or the loss from the body of HCO_3^-.

It should be noted that, in metabolic acidosis, there is often nothing wrong with the kidneys. They simply cannot excrete an H^+ load instantaneously. The respiratory compensation, by restoring pH towards normal, hampers the renal removal of the defect (i.e. the renal H$^+$ secretion and HCO$_3^-$ reabsorption/ regeneration). The exception to this is renal failure, where the kidneys' failure to excrete H$^+$ is the cause of metabolic acidosis (see Chapter 14).

10.4.5 *Metabolic alkalosis*

This is alkalosis which is not caused by a change in the arterial pCO$_2$. In the reaction sequence

$$CO_2 + H_2O \rightleftharpoons H_2CO_3 \rightleftharpoons H^+ + HCO_3^-$$
$$\searrow \qquad OH^-$$
$$H_2O \nearrow$$

we can regard metabolic alkalosis as the addition of base (OH$^-$ in the equation, although it can be any H$^+$ acceptor) to the system, thereby removing H$^+$ from the right of the reaction and so causing the reaction to move to the right, increasing plasma [HCO$_3^-$]. We can regard this as occurring initially at a constant (normal) pCO$_2$ (Figure 10.10).

$$\downarrow\uparrow pH \propto \frac{[HCO_3^-]\uparrow\uparrow}{pCO_2\uparrow}$$

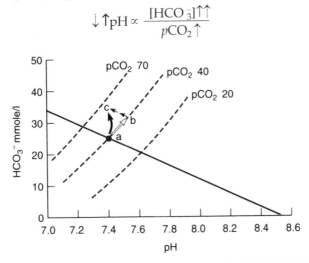

Figure 10.10 Metabolic alkalosis and its compensation. Metabolic alkalosis, the loss of H$^+$ or addition of base to the body (for any reason other than a change in pCO$_2$), can be regarded as occurring initially at a constant pCO$_2$. So as metabolic alkalosis develops, we move from point (a) to point (b) along the pCO$_2$=40 isobar. Plasma [HCO$_3^-$] increases (see text). The correction of the defect occurs as a result of a reduction in the rate of renal H$^+$ secretion. However, as this takes time to reduce the pH (and [HCO$_3^-$]), there is respiratory compensation. The increased pH, detected by the peripheral chemoreceptors, reduces alveolar ventilation and so increases the pCO$_2$ (and [HCO$_3^-$]). Thus we move from point (b) to point (c), parallel to the normal pH/HCO$_3^-$ line. As with metabolic acidosis, the onset of metabolic alkalosis brings the respiratory compensation into play immediately, so that the value of [HCO$_3^-$], pH and pCO$_2$ change as shown by the direct a–c arrow.

The decreased $[H^+]$, acting on chemoreceptors, reduces ventilation and so increases the pCO_2. This is **respiratory compensation for metabolic alkalosis.**

$$\downarrow\uparrow pH \propto \frac{[HCO_3^-]\uparrow\uparrow\uparrow}{pCO_2\uparrow\uparrow}$$

The compensation brings down the pH, but further increases the plasma $[HCO_3^-]$, although the effect of respiratory compensation in metabolic alkalosis is normally of very minor importance.

The removal of the defect is brought about by the effect of the defect on renal H^+ secretion. The defect is decreased $[H^+]$ in the body, including the renal tubule cells, so the rate of H^+ secretion is reduced. The filtered load of HCO_3^- is increased by the defect (and compensation), so that H^+ secretion is inadequate to reabsorb all the filtered HCO_3^- (or to generate more HCO_3^-). The plasma $[HCO_3^-]$ therefore falls, causing the reaction

$$CO_2 + H_2O \rightleftharpoons H_2CO_3 \rightleftharpoons H^+ + HCO_3^-$$

to move to the right, restoring $[H^+]$ to normal.

As in the case of metabolic acidosis, the respiratory compensation hampers the renal removal of the defect. This fact again emphasizes that it is the regulation of pH which is all important, rather than the regulation of $[HCO_3^-]$ or pCO_2 per se.

Causes of metabolic alkalosis
Metabolic alkalosis is most commonly caused by the loss of acid from the body by vomiting, thereby losing the hydrochloric acid from the stomach. Metabolic alkalosis is also caused by an increase of plasma bicarbonate. Changes in body fluid volume can alter plasma $[HCO_3^-]$, by altering the renal $[HCO_3^-]$ threshold.

Further reading

Burns K. D. and Levine, D. Z. (1997) Acid–base balance. In R. Wilkinson and Rex L. Jamison (eds), *Nephrology*, Chapman and Hall, London, pp. 117–133

Cogan, M. G. and Rector, F. C. Jnr. (1991) Acid–base disorders. In B. M. Brenner and F. C. Rector Jnr. (eds), *The kidney*, Saunders, Philadelphia, pp. 737–804

Good, D. W. and Knepper, N. A. (1990) Mechanism of ammonia excretion. Role of the renal medulla. *Semin. Nephrol.* **10**, 166–173

Problem

10.1   A subject is found to have an arterial pH of 7.25, a plasma $[HCO_3^-]$ of 14 mM and a pCO_2 of 33 mmHg. What acid–base disturbance is present?

Renal control of body fluid potassium content

11

11.1 Importance of K^+ in the body

In Chapter 8, the importance of Na^+ in the extracellular fluid was described and it was shown that the osmolality (of the extracellular fluid) depends on the body Na^+ content. Inside the cells, however, K^+ is the major cation and the maintenance of K^+ balance is essential for life. The body contains 3–4 mol K^+ (i.e. 120–160 g), but only about 2% of this is extracellular. The normal cellular concentration of K^+ is 150–160 mmol/l, whereas the normal plasma concentration is only 4–5 mmol/l. The causes of this uneven distribution were discussed in Chapter 1.

The ratio of intracellular $[K^+]$ to extracellular $[K^+]$ affects the resulting membrane potential of nerve and muscle cells, and hence affects their excitability. Hypokalaemia causes hyperpolarization of the nerve and muscle cells (i.e. the resting potential becomes more negative), making the cells less sensitive to depolarizing stimuli and therefore less excitable. Thus severe hypokalaemia can cause paralysis. Hyperkalaemia depolarizes cells, making them more excitable, but in severe hyperkalaemia the resting potential may be above (i.e. less negative than) the threshold potential, and this means that nerve cells are unable to repolarize after conducting an action potential, so that paralysis ensues.

Slow changes in the body K^+ content are much better tolerated than rapid changes. This is because, when the changes are slow, the equilibrium between intra- and extracellular potassium is maintained. Thus there may be hypokalaemia, but the ratio of intracellular $[K^+]$ to extracellular $[K^+]$ may be normal. In such circumstances, nerve and muscle cell excitability will also be normal.

11.2 *Regulation of body K^+*

A normal diet contains 40–120 mmol K^+ per day, more than enough to satisfy the needs of the body, and hence the maintenance of K^+ balance is brought about by regulation of K^+ excretion. However, since the vast majority of K^+ ions in the body are intracellular, these intracellular 'stores' tend to buffer any changes in plasma $[K^+]$, i.e. a small decrease in plasma $[K^+]$ causes some movement of K^+ out of cells and hence the change in plasma $[K^+]$ is minimized. This implies that, for K^+ control to be effective, very small changes in plasma $[K^+]$ must be able to provide the stimulus for excretion to be adjusted.

11.2.1 *Renal K^+ handling*

The normal dietary K^+ intake is almost equal to the Na^+ intake, but the plasma $[K^+]$, at 4 mmol/l, is less than 3% of the plasma concentration for Na^+ (140 mmol/l). Hence the filtered load of K^+ is much smaller than the filtered load of Na^+, so to maintain K^+ balance there must be a higher fractional excretion of K^+ than of Na^+.

In the proximal tubule, some 70% of filtered K^+ is reabsorbed, predominantly passively, and probably mainly through the tight junctions (i.e. paracellularly). However, proximal tubule cells can also reabsorb K^+ against a concentration gradient, so a component of the reabsorption appears to be active.

In the descending (thin) limb of the loop of Henle, there is some K^+ secretion. About 20% of filtered K^+ is reabsorbed from the ascending limb, together with Cl^- and Na^+, and although some of this K^+ immediately leaks back into the ascending limb, a proportion enters the medullary interstitium and hence the descending limb. Thus there is some K^+ recycling in the loop of Henle. (K^+ also recycles from the medullary collecting tubules to the loop of Henle, via the interstitium, in a manner similar to that of urea.)

In the early distal tubule, which is functionally similar to the thick ascending limb of Henle, Na^+, Cl^- and K^+ reabsorption occurs, but the rates of K^+ reabsorption and leakback into the nephron are similar, so there is little change in the tubular $[K^+]$. The late distal tubule and subsequent segments of the collecting-duct system secrete K^+ into the tubular fluid, and it is this secretion that accounts for the K^+ appearing in the urine. The secretory process is thought to be passive, driven by the electrochemical gradient between the cells and the lumen and occurs in the principal cells. Na^+ absorption to some extent determines the rate of K^+ secretion, but, in addition, any increase in the cellular $[K^+]$ will favour K^+ secretion and any decrease in cellular $[K^+]$ will reduce K^+ secretion. This provides a means of automatically adjusting K^+ secretion to the body's requirements for K^+ balance.

Changes in distal tubular lumen $[K^+]$ influence the rate of K^+ secretion. Increases in the tubular $[K^+]$ decrease the rate of secretion, whereas decreases in the tubular $[K^+]$ increase the rate of K^+ secretion. The $[K^+]$ in the distal tubule, produced by a given rate of distal K^+ secretion, will be determined by the flow rate in the distal tubule. Thus diuretics (see Chapter 15) which increase distal tubular flow can increase the rate of K^+ secretion by lowering the tubular K^+ concentration (as well as by increasing Na^+ delivery to the distal tubule).

The dependence of the K^+ secretory rate on the distal tubular flow rate also prevents or minimizes disturbances of K^+ balance in situations where the effective circulating volume is reduced. Reductions of the effective circulating volume increase aldosterone release (see Chapter 9), and this will tend to enhance distal nephron K^+ secretion (see below). However, decreased effective circulating volume also enhances sodium and water reabsorption in the proximal tubule, thereby decreasing the tubular fluid flow rate in the distal nephron (reductions in glomerular filtration rate may also occur, with a similar effect on distal tubular flow). The reduced distal tubular flow will decrease K^+ secretion and counteract the stimulatory effect of aldosterone on K^+ secretion.

ADH stimulates potassium secretion across the apical membrane of the collecting tubule cells, by enhancing Na^+ permeability so that more Na^+ enters the cells from the tubular lumen, making the lumen potential more negative. This effect of ADH serves to prevent changes in urine volume from disturbing K^+ homeostasis. Since K^+ secretion in the distal nephron is reduced at low tubular flow rates, ADH-induced antidiuresis would otherwise be accompanied by K^+ retention, and diuresis would be accompanied by K^+ loss.

11.2.2 *Aldosterone*

Since the K^+ losing effects of aldosterone (p. 109) do not exhibit the escape phenomenon' in the way that the Na^+-retaining effects do, there are grounds for considering that, normally, aldosterone plays a more important role in determining K^+ balance than in regulating Na^+ excretion.

In the distal nephron, the aldosterone-sensitive transport mechanisms for sodium and potassium are, to a large extent independent of each other. The 'escape phenomenon' mentioned above is itself strong evidence for independent mechanisms for Na^+ and K^+, but, in addition, the drug actinomycin D blocks the Na^+-retaining effects of aldosterone without inhibiting the K^+-losing effect, and, in adrenalectomized animals, very small amounts of aldosterone restore the plasma Na^+ concentration to normal, but do not significantly lower the elevated plasma K^+ concentration. This finding emphasizes that, whereas for Na^+, aldosterone has a 'permissive action' (see Chapter 9), for K^+, aldosterone has a regulatory function.

Increases in plasma $[K^+]$ act directly on the adrenal cortex to increase aldosterone output, and decreases in plasma $[K^+]$ reduce aldosterone output. Aldosterone constitutes essentially the only hormonal control over K^+ output, whereas it is only one of many factors regulating Na^+ output.

11.3 *Hypokalaemia*

Since the diet almost invariably contains adequate K^+, hypokalaemia is generally due to losses of K^+ from the gastrointestinal tract or by the kidneys. Persistent vomiting or diarrhoea, or the use of certain commonly prescribed diuretics (Chapter 15), are frequent causes of hypokalaemia.

The hypokalaemia in diarrhoea is due to the faecal loss of K^+ from the gastrointestinal secretions. Vomiting results in loss of some K^+ directly in the vomitus, but the main effect of vomiting on K^+ balance is due to a change in urinary K^+ excretion. Vomiting causes metabolic alkalosis (due to the loss from the body of gastric HCl), and alkalosis reduces proximal tubular HCO_3^- absorption (Chapter 10), and also reduces proximal Na^+ absorption (since HCO_3^- stays in the tubule, the associated cation, Na^+, also stays in the tubule) and water absorption. The increased $NaHCO_3$ delivery to the distal tubule enhances Na^+ absorption at this site, and H^+ and K^+ are secreted in increased amounts. The increased tubular flow rate facilitates this. In addition, the loss of NaCl and volume depletion brought about by vomiting tends to increase aldosterone release from the adrenal cortex, which exacerbates the urinary K^+ loss.

Another relatively common cause of hypokalaemia is excess insulin (either exogenous insulin during diabetic treatment, or endogenous insulin). Insulin increases K^+ entry into cells (of skeletal muscle and liver), so that although the total amount of K^+ in the body is not altered directly, the extracellular $[K^+]$ decreases.

The physiological effects of a particular degree of hypokalaemia are widely different in different subjects. Most subjects remain symptom-free until plasma $[K^+]$ has fallen to approximately one-half its normal value (i.e. down to 2–2.5 mmol/l). The initial symptom is muscle weakness, usually affecting the lower extremities and gradually extending upwards, until death occurs when respiratory function becomes affected.

Potassium deficiency causes numerous other derangements of metabolism. The synthesis of liver and muscle glycogen requires potassium, so, since the conversion of glucose into glycogen is altered by hypokalaemia, the condition produces an abnormal glucose tolerance test. Hypokalaemia also affects vascular tone (causes vasoconstriction). Polyuria and thirst are present, because the renal response to ADH is impaired by hypokalaemia, so patients are unable to produce concentrated urine (see below). Metabolic alkalosis is frequently present in hypokalaemic patients, since the K^+ deficit tends to cause an increased intracellular $[H^+]$, which leads, in the distal tubule cells, to increased H^+ secretion. As the physiological effects of a particular degree of hypokalaemia are widely different in different subjects, it is best to assess the functional effects of hypokalaemia. This can be done by monitoring the electrocardiogram (ECG) and muscle strength.

Cardiac muscle is an excitable tissue dependent on K^+ for its normal functioning. The K^+ permeability of the cardiac muscle cell membrane varies

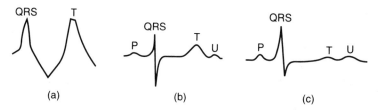

Figure 11.1 Changes in the electrocardiogram (ECG) in disorders of potassium homeostasis (lead aVL recording). (a) Hyperkalaemia, with characteristic peaked T wave and widening of QRS complex (plasma $[K^+]=8$ mM). (b) Normal ECG (plasma $[K^+]=4$ mM). (c) Hypokalaemia, showing characteristic flattened T wave and ST segment depression. The U wave is often prominent and may be mistaken for the T wave (plasma $[K^+]=2$ mM).

directly with plasma $[K^+]$. After excitation, the repolarization of the muscle is brought about by an increase in K^+ permeability, causing K^+ to move out of the cells. In hypokalaemia, the time for cardiac muscle to repolarize is prolonged. The effect of this on the ECG is shown in Figure 11.1. These characteristic ECG changes appear when plasma K^+ concentration falls to about 3 mmol/l. There is ST segment depression, decreased amplitude (or inversion) of the T wave, a much enlarged U wave and (frequently) arrhythmias. (The arrhythmias of hypokalaemia are more severe in subjects taking digitalis.)

Treatment consists of the oral or intravenous administration of a potassium salt. This must be done with great care, however, since hyperkalaemia can be readily produced. Continual monitoring of the ECG is the best way of monitoring the effect of the K^+ administration. It is difficult to predict the extent of a K^+ deficit, since the plasma $[K^+]$ is a poor guide to K^+ depletion. In fact, because acidosis causes the release of K^- from cells, it is possible for the plasma $[K^+]$ to be elevated in a patient who is K^+ depleted. Similarly, in alkalosis, the entry of K^+ into cells can depress the plasma K^+ even in subjects with excess total K^+. Thus assessment of the K^+-balance status of a patient cannot readily be made unless any acid–base disturbances are first corrected.

There are a number of ways in which K^+ can be replaced in potassium deficit states. Preparations for oral or intravenous administration include KCl, $KHCO_3$ and K_2HPO_4. There are advantages in using KCl, because many K^+-depleted patients are also Cl-deficient (e.g. because of diuretics or vomiting) and have metabolic alkalosis. In a minority of patients who are K^+ depleted and also have metabolic acidosis, $KHCO_3$ is the most appropriate K^+ salt to administer.

11.3.1 *Renal function in hypokalaemia*

At the onset of hypokalaemia (e.g. if the diet is changed to a K^+-deficient one), the kidney does not immediately conserve K^+ effectively and urinary K^+ output remains greater than about 30 mmol/day for 2–3 weeks. During this period, the kidney becomes progressively more efficient at K^+ conservation and K^+ output falls below 30 mmol/day. Thus hypokalaemia and a low urinary K^+ output is indicative of long-standing K^+ depletion, caused by extrarenal factors (e.g. gastrointestinal loss). Hypokalaemia with a normal or high K^+ output indicates that the K^+ depletion has occurred recently, or that the kidney is the *cause* of the hypokalaemia.

The polyuria and inability to produce concentrated urine, which are usually present in hypokalaemic subjects, are due to:

(1) diminished medullary concentration gradient;
(2) resistance of the collecting ducts to the effects of ADH;
(3) polydipsia.

Potassium depletion also causes some anatomical alterations within the kidney. In the proximal tubule, the cells may develop numerous vacuoles, and in the interstitial spaces fibrosis may occur. Most of these defects disappear when normokalaemia is restored.

Severe hypokalaemia also alters the collecting tubule handling of K^+. There is a fall in the luminal membrane K^+ permeability (possibly due to a decreased plasma aldosterone level), which reduces K^+ secretion. In this situation a K^+ reabsorptive process is apparent, dependent on the Na^+ gradient (probably $Na^+/K^+/2Cl^-$ reabsorption).

11.4 *Hyperkalaemia*

Excess K^+ is normally removed from the body by renal secretion. Ingestion of excess K^+ causes a small rise in plasma $[K^+]$ ('buffered' by the entry of K^+ into cells) and the increased plasma $[K^+]$ stimulates the release of aldosterone from the adrenal cortex, which promotes K^+ secretion. A high K^+ intake can increase the number of K^+ channels in the apical membranes of the distal nephron principal cells, independently of any change in aldosterone levels. The mechanism of this is currently unknown.

It is rare for excess input of K^+ to present problems. Normal subjects can tolerate tenfold increases in K^+ intake (such as the transfusion of stored blood, which may contain 30 mM K^+). However, patients with impaired kidney function, or infants, may suffer hyperkalaemia which can be fatal.

As mentioned above, acidosis can cause hyperkalaemia even when the body's K^+ stores are normal. Insulin promotes K^+ entry into cells, so insulin deficiency can lead to hyperkalaemia. Another cause of hyperkalaemia is the excessive breakdown of cells, e.g. after severe trauma, or treatment with cytotoxic drugs.

Hyperkalaemia can occur as a result of decreased K^+ excretion; in renal failure, as the number of effective nephrons decreases, K^+ excretion per functioning nephron increases, so that K^+ balance can be maintained. However, when renal failure reaches the stage where a reduction in urine flow occurs, K^+ excretion falls and hyperkalaemia develops. The reduction in the kidney's ability to excrete K^+ in oliguric renal failure is probably due to decreased fluid (and Na^+) delivery to the distal K^+ secreting site.

The reduced intracellular:extracellular $[K^+]$ ratio which occurs in hyperkalaemia, decreases (i.e. makes less negative) the potential across cell membranes and, in excitable cells (nerve and muscle), if the depolarization reaches the threshold, the cells are unable to conduct further action potentials and muscle weakness (and in extreme cases, paralysis) ensues. This is a characteristic feature of hyperkalaemia, as are ECG changes (Figure 11.1). The use of K^+ sparing diuretics (see Chapter 15), or the presence of renal failure, may exacerbate hyperkalaemia.

Several methods of treatment are possible. Loop diuretics can be used (Chapter 15) to promote K^+ excretion, or insulin (or glucose, which increases insulin release) can be administered to promote K^+ entry into cells.

The effects of hyperkalaemia on muscle function can be corrected even in the continuing presence of hyperkalaemia, by administering Ca^{2+}. This makes a larger depolarization necessary to reach threshold, so can return cell excitability towards normal. The effects of Ca^{2+} are transient, so it can be used as a short-term measure while waiting for insulin, for example, to take effect.

Further reading

Giebisch, G. (1998) Renal potassium transport: mechanisms and regulation. *Am. J. Physiol.* **274**, F817–F833

Wang, W., Hebert, S. C. and Giebisch, G. (1997) Renal K^+ channels: structure and function. *Annu. Rev. Physiol.* **59**, 413–436

Renal regulation of *12*
body calcium, magnesium
and phosphate

12.1 Introduction

Calcium is the most abundant cation in the body, there being about 25 mol (1 kg) in an average 70 kg man. Almost all of this calcium is within bone, which consists essentially of complex salts of calcium and phosphate. However, both calcium and phosphorus (as phosphate) have important extraskeletal functions. Magnesium has important intracellular actions, as well as being a constituent of bone.

12.2 Calcium

Precise control of the extracellular fluid (including plasma) calcium concentration is necessary because of the effects of calcium on excitable tissues (nerve and muscle). The excitability of nerve and muscle cell membranes depends on the difference between the resting membrane potential and the threshold potential. The threshold potential varies inversely with the plasma Ca^{2+} concentration.

Calcium is present in two forms in the plasma:

(1) ionized calcium, Ca^{2+}, normal concentration about 1.25 mmol/l (5 mg/100 ml);
(2) bound calcium, mainly bound to protein (especially albumin) but some complexed with organic acids, the concentration is also about 1.25 mmol/l.

Physiologically, the ionized calcium is more important, but generally measurements of plasma calcium are of *total* calcium, for which the normal value is about 2.50 mmol/l. It should be noted that this value is not for the ECF as a whole, since the interstitial fluid will have very little bound calcium (there being only a small amount of protein in the interstitial fluid), so interstitial fluid total calcium is closer to 1.25 mmol/l.

The intracellular calcium is mostly sequestered in organelles (smooth endoplasmic reticulum, mitochondria), or complexed with macromolecules (including specific calcium-binding proteins such as calmodulin). Hence the intracellular concentration of ionized calcium is very low (0.0001 mmol/l, compared with the extracellular fluid value of 1.25 mmol/1). Active mechanisms must be involved in maintaining this low intracellular Ca^{2+} concentration.

12.2.1 *Maintenance of calcium balance*

Figure 12.1 shows the normal dietary intake and faecal and urinary losses of calcium. Normally, the urinary loss per day is equal to the net intestinal absorption.

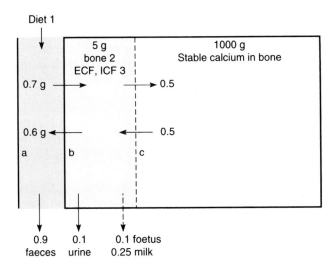

Figure 12.1 Daily calcium intake and output. All numerical values in the diagram are grams. Functionally, the calcium in the body can be divided into three pools, labelled a, b, c.

(a) The gut. This normally receives 1 g per day (25 mmol) in the diet, and about 700 mg (17.5 mmol) of this is absorbed. However, the net reabsorption is only 100 mg (2.5 mmol), since 600 mg (15 mmol) re-enters the gut. Thus the faeces contain 900 mg (22.5 mmol) calcium per day.

(b) The 'pool' of exchangeable calcium. About 5 g calcium in the body is 'exchangeable'. This is present on bone surfaces (2 g, 50 mmol) and in the body fluids (3 g, 75 mmol). Exchanges occur between this compartment and the gut, and also between this compartment and the 'stable' calcium.

(c) Stable calcium in bone. There is about 1000 g (25 mol) of this, but only 0.5 g (12.5 mmol) exchanges daily with the 'exchangeable' calcium pool. Losses of calcium from the exchangeable pool occur normally in the urine. During pregnancy and lactation there are additional requirements for calcium, and net gut absorption increases.

Table 12.1 Fractional Ca^{2+} and Mg^{2+} reabsorption along the nephron

	Filtered load reabsorbed (%)	
	Ca^{2+}	Mg^{2+}
Proximal tubule (pars convoluta)	60	15
Proximal tubule (pars recta)	10	15
Thick ascending limb of Henle	20–25	60
Distal convoluted tubule	5–10	2–5
Collecting tubule system	<0.5	<0.5

12.2.2 *Calcium handling by the kidney*

Approximately 50% of the plasma calcium is bound to plasma protein and is therefore not filterable. The remainder (mainly ionized calcium) is filtered, but normally less than 5% of the filtered calcium appears in the final urine. Since the plasma ionized calcium concentration is 1.25 mmol/l, the amount filtered per day is 1.25×180 mmol, i.e. 225 mmol, most of which is reabsorbed (Table 12.1).

Proximal tubule
Calcium reabsorption in the proximal tubule parallels the reabsorption of sodium and water, and the calcium concentration therefore stays approximately constant along the proximal tubule.

Since Ca^{2+} is positively charged, there are, energetically, no problems in the entry of calcium into the cell but the membrane calcium permeability is extremely low. However, at the peritubular side, the nature of the transport process is still not clear. A calcium-activated ATPase has been demonstrated in the peritubular cell membrane, but, in addition, Ca^{2+} counter-transport out of the cell, coupled to passive Na^+ entry, occurs, the energy being derived from active sodium transport which maintains a low intracellular Na^+ concentration (the ratio appears to be $3Na^+$ entering for $1Ca^{2+}$ leaving).

Loop of Henle
In the ascending limb of the loop of Henle, calcium reabsorption occurs and, as in the proximal tubule, Ca^{2+} reabsorption is mainly passive. Furosemide, a diuretic which acts primarily by inhibiting NaCl transport in the loop of Henle, also inhibits calcium reabsorption, probably by reducing the positive potential in the tubular lumen, and so decreasing paracellular Ca^{2+} reabsorption.

Distal tubule and collecting duct
Normally, 10–12% of the filtered load of calcium is delivered to the distal tubule (this is the same proportion as for sodium) and about two-thirds of this calcium is reabsorbed. The lumen of the tubule in this part of the nephron is

negative (pp. 73–74), so that active calcium absorption is thought to occur, against the electrochemical gradient. The physiological regulation of Ca^{2+} reabsorption occurs mainly in the cortical thick ascending limb and the distal tubule.

Table 12.1 shows the fractions of filtered Ca^{2+} reabsorbed along the nephron.

12.3 *Phosphate*

Inorganic phosphate exists in the plasma and interstitial fluid in two forms, 'acid' phosphate, $H_2PO_4^-$, and 'alkaline' phosphate, HPO_4^{2-}. A third form, PO_4^{3-}, can exist, but does not occur at physiological pH. Generally the term 'phosphate' is used to refer to all of these forms.

Inside the cells, phosphate is present not only as 'acid' and 'alkaline' inorganic phosphate, but also in organic molecules such as ATP, ADP and cyclic AMP. In the extracellular fluid, the proportions of the two forms of inorganic phosphate are determined by the pH. The pK for the interconversion of the forms,

$$H_2PO_4^- \rightleftharpoons H^+ + HPO_4^{2-}$$

is 6.8, i.e. at pH 6.8 $[HPO_4^{2-}]=[H_2PO_4^-]$. However, since the normal plasma pH is 7.4, there will be more HPO_4^{2-} than $H_2PO_4^-$, the ratio being $4:1$.

This can be derived from the Henderson–Hasselbalch equation (Chapter 10) where

$$pH = pK + \log \frac{[base]}{[acid]}$$

so for phosphate

$$7.4 = 6.8 + \log \frac{[HPO_4^{2-}]}{[H_2PO_4^-]}$$

$$\log \frac{[HPO_4^{2-}]}{[H_2PO_4^-]} = 0.6$$

$$\therefore \frac{[HPO_4^{2-}]}{[H_2PO_4^-]} = \frac{4}{1}$$

12.3.1 *Maintenance of phosphate balance*

Phosphate is filtered at the glomerulus into the nephron, so that the $4:1$ ratio of alkaline to acid phosphate is present in the glomerular filtrate. Conversion of alkaline to acid phosphate occurs in the tubules as a result of H^+ secretion (Chapter 10).

The phosphate present in the plasma is normally expressed as an amount of elemental phosphorus (P). The normal range is 0.8–1.3 mmol/l. At a concentration of 1.0 mmol/l phosphorus, almost all (95%) of the filtered phosphate is reabsorbed. This reabsorption mainly occurs in the early proximal tubule.

The only hormone which is definitely known to regulate renal tubular phosphate transport physiologically is PTH (parathyroid hormone), although other peptide hormones – notably calcitonin, glucagon and insulin – may also influence renal phosphate transport. Generally, PTH, calcitonin and glucagon increase renal phosphate excretion, whereas insulin reduces phosphate excretion.

12.4 *Calcium and phosphate homeostasis (Figure 12.1)*

Calcium and phosphate can enter the ECF from:

(1) the intestine (via dietary intake);
(2) bone stores;

and can leave the ECF:

(1) via the kidneys (in urine);
(2) into bone.

There is an inverse relationship between plasma calcium and phosphate concentrations, because $[Ca^{2+}] \times [phosphate]$ is close to the solubility product, so an increase in $[Ca^{2+}]$ will cause the precipitation of calcium phosphate (in bone), thus lowering the phosphate concentration. Similarly an increase in phosphate concentration lowers the calcium concentration. But *small* changes in plasma calcium may lead to *parallel* (rather than inverse) changes in plasma phosphate. Nevertheless, the mechanisms whereby calcium and phosphate are regulated are closely interrelated. The two major regulators of calcium and phosphate in the body are parathyroid hormone (PTH) and vitamin D.

Parathyroid hormone is a polypeptide secreted from the parathyroid glands. Secretion is stimulated by a decrease in the plasma ionized calcium concentration, and reduced by an increase in the plasma ionized calcium concentration. Changes in phosphate concentration also change PTH secretion, but this may occur as a result of consequent changes in Ca^{2+}. Figure 12.2 shows the main actions of PTH.

Vitamin D is a steroid. It is derived from precursors which can be either ingested in the diet or produced by ultraviolet light acting on the skin. The active hormone, 1,25-dihydroxycholecalciferol, also termed calcitriol, is produced by a series of metabolic steps in the liver and kidneys, as shown in Figure 12.3. Essentially its action is to increase the availability of calcium and phosphate by enhancing their intestinal absorption and their release from bone (in the presence of PTH), but it may also reduce the urinary excretion of

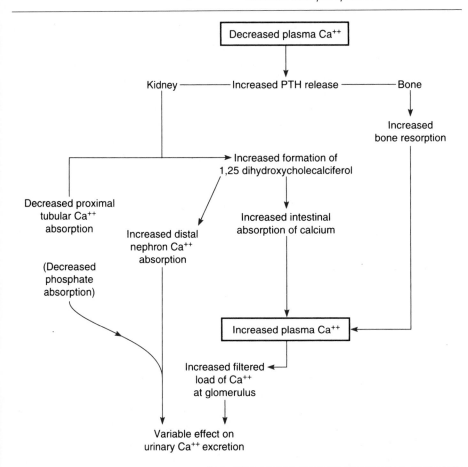

Figure 12.2 The actions of parathyroid hormone (PTH).

calcium and phosphate (Figure 12.3). Whereas the single most important action of PTH is on bone, that of vitamin D is on the intestine.

There is a third humoral agent involved in the control of calcium phosphate: calcitonin, a peptide hormone which is produced by the parafollicular cells of the thyroid gland. Its effect is to decrease the extracellular fluid calcium concentration by reducing the release of calcium from bone.

12.4.1 *Disturbances of calcium and phosphate homeostasis*

Hypocalcaemia

The characteristic feature of a low plasma $[Ca^{2+}]$ is tetany – convulsions and muscle cramps involving the hands and feet. For nerve and muscle excitability,

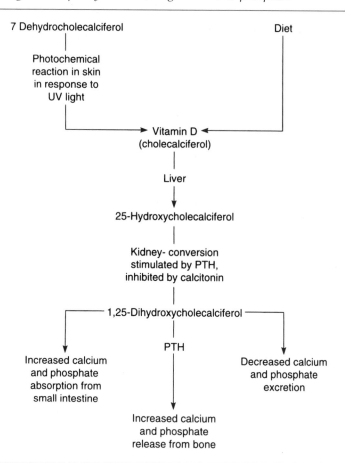

Figure 12.3   The actions of vitamin D.

it is the ionized calcium which is important, so it is possible for a severe disturbance to be present even if the total plasma calcium is normal. Such disturbances can occur as a result of acidosis or alkalosis. Normally, as we have seen above, the plasma has equal concentrations of bound and ionized calcium. The bound calcium depends on electrostatic charge for its binding (the binding is to the negative charges on the protein molecules). Acidosis causes increased H^+ binding to the negative binding sites on the proteins, so less Ca^{2+} can be bound, and free $[Ca^{2+}]$ rises. Alkalosis has the opposite effect, leading to a fall in free $[Ca^{2+}]$, so that the symptoms of hypocalcaemia appear even though the total plasma calcium is normal.

Conversely, it is possible for the total plasma calcium to be low, without any symptoms of hypocalcaemia. This occurs in hypoalbuminaemia, where the regulatory mechanisms (PTH, vitamin D, calcitonin) keep the plasma ionized calcium concentration normal, but, because the amount of albumin is reduced, the amount of calcium bound to the albumin is also reduced.

Hypocalcaemia is a common finding in patients with renal failure. It is apparently due mainly to hyperphosphataemia and bone resistance to PTH. The amount of phosphate excreted is the difference between the amount filtered and the amount reabsorbed. In renal failure the amount reabsorbed is often normal, but GFR is reduced and hence phosphate excretion is diminished. The plasma phosphate concentration rises, and since $[Ca^{2+}] \times$ [phosphate] = constant in this situation (see above) there is hypocalcaemia. (This occurs as a result of the deposition of calcium phosphate salts in bone.) Patients with severe renal failure may be unable to increase phosphate excretion sufficiently to lower the plasma phosphate concentration (this is true for GFR values of less than 30 ml/min) and deposition of calcium and phosphate may then occur in sites other than the skeleton (e.g. muscle, blood vessels, the heart). This is termed metastatic calcification.

Hypercalcaemia

This can occur as a result of sudden acidosis (see above), releasing 'bound' calcium to become ionized Ca^{2+}, so that symptoms of hypercalcaemia occur even though total plasma calcium is normal. True hypercalcaemia (increases in both ionized and bound calcium) can occur as a result of increased intestinal absorption (due to excess vitamin D), or increased release of calcium from bone due to bone disease, or excess PTH, or substances with actions like PTH released from tumours (such as carcinoma of the bronchus).

A major symptom of hypercalcaemia is renal calculi, but in addition there may be disturbances of behaviour (due to effects on higher cerebral function), disturbed intestinal motility and renal damage (due to a toxic effect of calcium on the renal tubules). There may be calcification in extraskeletal sites.

12.5 *Magnesium*

Magnesium is quantitatively the second most important intracellular cation (K^+ being the most important). The importance of magnesium stems from its role in energy storage and in the control of mitochondrial oxidative metabolism. Magnesium-ATP regulates energy production by mitochondria, and magnesium is also vital for protein synthesis. In addition, it plays a part in regulating K^+ and Ca^{2+} channels in cell membranes.

The distribution of magnesium in the body is: bone 55%, intracellular fluid 44%, plasma 0.4%, other extracellular fluids 0.6% and the total body Mg content is about 28 g. The fraction of plasma magnesium which is filterable by the kidneys is 75%, most of which is ionized magnesium, Mg^{2+}. There is a small fraction of magnesium which is filterable but unionized, complexed to citrate, oxalate and phosphate, and to a lesser extent to sulphate and bicarbonate.

Of the plasma magnesium 25% is protein bound and hence is not filterable. This fraction is not present in interstitial fluid. Thus, whereas the plasma total magnesium concentration is about 1.5–2.0 mmol/l, the interstitial fluid magnesium concentration is 1.1–1.5 mmol/l.

12.5.1 *Magnesium homeostasis*

Figure 12.4 shows the daily intake and output of magnesium. The gastrointestinal absorption and urinary excretion of magnesium are normally equal.

12.5.2 *Renal reabsorption of magnesium*

Although there is some magnesium reabsorption in the proximal tubule (Table 12.1), the permeability of the epithelium is much lower for Mg^{2+} than for Ca^{2+} or Na^+. Consequently, the tubular fluid Mg^{2+} concentration increases about 1.5-fold along the proximal tubule.

The main site at which the filtered Mg^{2+} is absorbed is the thick ascending limb of the loop of Henle. Much of this reabsorption of Mg^{2+} is passive and mainly paracellular (between the cells), driven by the lumen-positive potential caused by the excess of Cl^- transport over Na^+ with back-diffusion of K^+ (Figure 12.5). However, there is also some passive entry of Mg^{2+} into the cells

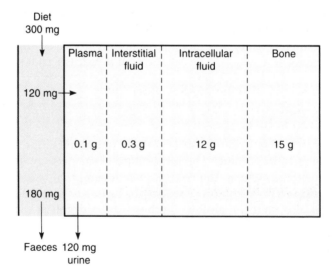

Figure 12.4 Daily magnesium intake and output. The main site of magnesium absorption is the small intestine. The percentage of magnesium absorbed decreases with increasing dietary intake.

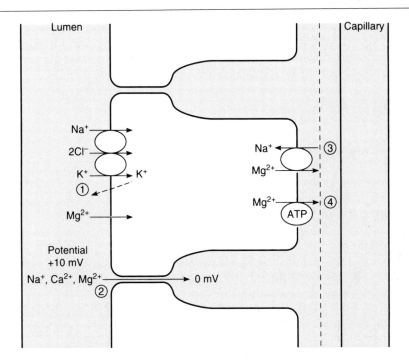

Figure 12.5 Magnesium absorption in the thick ascending limb of the loop of Henle. The $Na^+/K^+/2Cl^-$ cotransporter in the apical membrane, with backleak of some K^+, ①, establishes a lumen-positive potential which drives Mg^{2+} absorption by the paracellular route. ② Ca^{2+} and Na^+ are also absorbed in this way. There is also some Na^+/Mg^{2+} antiport ③ across the basal membrane, as well as Mg^{2+}-ATPase ④.

from the lumen, and this Mg^{2+} is then transported across the basal cell membrane by the Mg^{2+}-ATPase and by Na^+–Mg^{2+} exchange (counter transport), as shown in Figure 12.5.

There is a little Mg^{2+} reabsorption by the distal tubule and collecting duct (see table 12.1).

12.5.3 *Regulation of magnesium balance*

For the kidney as a whole, magnesium behaves as if there is a T_m for magnesium reabsorption, with the normal T_m equal to the normal filtered load of magnesium. Consequently, small increases in the plasma magnesium concentrations, which increase the filtered load, exceed the T_m and hence increased amounts of magnesium are excreted.

There is some evidence for intrinsic regulation of magnesium reabsorption by cells of the thick ascending limb of the loop of Henle. If the magnesium concentration of the fluid in contact with the cells is low (e.g. because of a fall in plasma [Mg^{2+}] and hence a reduced filtered Mg load), the cells transport more Mg^{2+}. However, magnesium reabsorption is also under hormonal control.

Parathyroid hormone (PTH) increases the tubular reabsorption of magnesium, and so diminishes magnesium excretion. This effect occurs in the thick ascending limb of the loop of Henle.

12.5.4   *Disorders of magnesium balance*

Gastrointestinal disorders of magnesium balance include decreased intake, diarrhoeal states and selective magnesium absorption disorders. Disorders of fat absorption also impair magnesium absorption. In the presence of any of these conditions, renal magnesium reabsorption is normally increased. However, the kidneys can also be the cause of magnesium depletion. A selective defect in renal magnesium reabsorption does occur, but in general renal magnesium losses are due to drugs. Most diuretics (except acetazolamide) increase magnesium excretion, as does alcohol (ethanol).

There are interactions between calcium and magnesium metabolism which are not fully understood. Magnesium depletion leads to hypocalcaemia, but neither PTH nor vitamin D appear to be involved in this interaction.

Further reading

Kurokawa, K. (1996) How is the plasma calcium held constant? *Kidney Int.* **49**, 1760–1764

Lote, C. J. (1997) Divalent ions. In R. L. Jamison and R. Wilkinson (eds), *Nephrology*. Chapman and Hall, London, pp. 109–116

Pazianas, M. and Eastwood, J. B. (1997) Calcium and phosphate metabolism. In R. L. Jamison and R. Wilkinson (eds), *Nephrology*. Chapman and Hall, London, pp. 358–368

Tenenhouse, H. S. (1997) Cellular and molecular mechanisms of renal phosphate transport. *J. Bone Mineral Res.* **12**, 159–164

Summary of the principal reabsorptive and secretory processes in the nephron segments

13

13.1 *Introduction*

In the foregoing chapters, the major ways in which solutes and water are reabsorbed and secreted along the nephron are reviewed. In this chapter, the data are summarized. However, although an attempt has been made to relate the amounts of sodium and water absorbed to the GFR in humans, the data for the segmental absorptions are determined by micropuncture techniques which cannot be used in man and the figures presented must therefore be regarded as approximations.

13.2 *Sodium*

Table 13.1 shows the extent of sodium reabsorption along the nephron. Changes in sodium reabsorption in the collecting duct provide the fine control of sodium excretion in normal circumstances. It is likely that glomerulo-tubular balance keeps the delivery of sodium to the collecting ducts approximately constant.

13.3 *Water*

The pattern of water reabsorption along the nephron has some important differences from that of sodium. In the presence of normal amounts of plasma

Table 13.1 Segmental sodium reabsorption

Segment	Approx. glomerular filtrate reabsorbed (%)	Amount Na reabsorbed/day (mmol)
Proximal tubule	70	17 800
Loop of Henle	20	5 200
'Distal tubule' and		
collecting ducts	9	2 200
Total	99	25 200

Table 13.2 Segmental water reabsorption

Segment	Glomerular filtrate reabsorption (%)	
	Normal ADH present	ADH absent
Proximal tubule	70	70
Loop of Henle	15	14
'Distal tubule' and		
collecting ducts	14	3
Total	99	87
Volume excreted (l/day)	1.5	23

ADH (such that urine volume is about 1.5 l/day), it can be seen (Table 13.2) that there is water reabsorption along with Na^+ in the proximal tubule, but that there is less water reabsorption (as a percentage of the amount filtered) than Na^+ absorption in the loop of Henle. This is because NaCl extrusion from the ascending limb of Henle is not accompanied by H_2O reabsorption (the ascending limb is impermeable to water). The water reabsorption occurring in the loop of Henle is mainly from the descending limb into the hypertonic medullary interstitium.

The water permeability of the early distal tubule is low and is not affected by changes in plasma ADH level (i.e. it has similar properties to the thick ascending limb of the loop of Henle). The later parts of the anatomical distal tubule behave like the collecting tubules, with ADH controlling water permeability. So, for the distal tubules as a whole, water reabsorption occurs to a greater extent than Na^+ reabsorption, because the fluid entering the distal tubule is hypotonic to plasma (as a result of the NaCl extrusion in the ascending limb of Henle). There is thus an osmotic gradient for distal tubular water reabsorption, which is therefore largely independent of local sodium reabsorption.

In the collecting ducts, water reabsorption is dependent on ADH; much of the water delivered to the collecting ducts is reabsorbed in the cortical part of the duct and the tubular fluid osmolality here increases from 100 mosmol/kg H$_2$O to almost 300 mosmol/kg H$_2$O. In the medullary collecting ducts, water reabsorption continues, leading to the excretion of hypertonic urine.

In the absence of ADH, water reabsorption is modified, not only in the collecting ducts, but also (to a lesser extent) in the loop of Henle. The impermeability of the collecting ducts to water (including the cortical collecting ducts) causes the excretion of a large volume (23 l/day) of hypotonic urine. Since Na$^+$ reabsorption occurs in the collecting ducts, urine osmolality can be as low as 60 mosmol/kg H$_2$O. An additional consequence of the absence of ADH is that the medullary tissue osmolality is reduced (due mainly to the washout of urea by the vasa recta and the fact that the rate of urea entry to the medulla from the collecting ducts is reduced). This reduced medullary interstitial osmolality decreases the gradient for water reabsorption from the descending limb of the loop of Henle, so that reabsorption in this segment is somewhat reduced.

13.4 *Potassium*

About 70% of the filtered potassium is reabsorbed in the proximal tubule, and reabsorption continues in the ascending limb of the loop of Henle. Most of the potassium which appears in the urine arrives there by secretion from the distal tubule. The collecting ducts are, however, also capable of transporting K$^+$, either reabsorbing it or secreting it depending upon the requirements of the body.

13.5 *Hydrogen ions and HCO$_3^-$*

H$^+$ secretion occurs in the proximal tubule (mainly by exchange for Na$^+$), the distal tubule (active) and the collecting tubule/ducts (active). In the proximal tubule, H$^+$ secretion is responsible for HCO$_3^-$ reabsorption, and there is only a 'small' reduction in tubular fluid pH (from 7.4 down to approximately 7.0). About 10% of the filtered HCO$_3^-$ is still unreabsorbed by the beginning of the distal tubule. H$^+$ secretion leads to the reabsorption of almost all of this, although there is little carbonic anhydrase activity in the lumen of the distal tubule. Alkaline phosphate is also converted into acid phosphate by tubular H$^+$ secretion (mainly distal), and the tubular fluid pH decreases. The biggest fall in pH occurs in the collecting ducts, and urinary pH can be as low as 4.5. This demands the maintenance of a very steep plasma-to-lumen H$^+$ gradient by the collecting duct cells (an [H$^+$] 1000 times greater in the lumen than in the plasma) and active H$^+$ secretion is essential for this.

13.6 *Phosphate*

Ninety per cent of filtered phosphate is reabsorbed in the proximal tubule, by active or secondary active transport. Parathyroid hormone inhibits phosphate reabsorption.

13.7 *Calcium*

Only 50% of plasma calcium is filterable; the rest is bound to proteins, etc., and is unable to cross the glomerular filter. The filtered calcium (Ca^{2+}) is reabsorbed throughout the nephron and this reabsorption appears in some parts of the nephron (proximal tubule, medullary ascending limb of Henle) to be linked to NaCl reabsorption.

Parathyroid hormone (PTH) affects calcium reabsorption. It decreases proximal calcium ion reabsorption, but increases reabsorption in the distal nephron, so that overall it increases tubular calcium ion reabsorption. This tubular action of PTH is usually obscured by the fact that PTH increases the plasma calcium concentration by increasing bone resorption, so that the increased filtered load of calcium may increase calcium excretion.

13.8 *Glucose*

At physiological plasma glucose concentrations, all of the filtered glucose is reabsorbed in the proximal tubule. Only if the plasma glucose concentration rises dramatically is there glucose excretion in the urine. Recent evidence indicates that some glucose reabsorption can occur in more distal parts of the nephron.

13.9 *Urea and other nitrogenous waste products*

Proximal tubular water reabsorption increases the luminal urea concentration, so that urea (which can readily cross most cell membranes) is passively reabsorbed. In the cortical collecting ducts, urea is unable to escape from the tubular lumen and consequently, if water is being absorbed (i.e. ADH present), the luminal urea concentration rises. Then, in the medullary collecting ducts (if ADH is present) urea is reabsorbed down its concentration gradient, to achieve a high concentration in the medullary interstitial fluid. This high interstitial concentration causes urea to enter the descending limb of Henle, so that there is some recycling of urea from interstitium to descending limb to collecting duct to interstitium.

Urea is highly soluble in water and has a very low toxicity. Hence precise control over its plasma concentration is not necessary and the plasma concentration normally varies over a threefold range (2.5–7.5 mmol/l). The high permeability of cell membranes to urea ensures that such changes in concentration do not alter the distribution of water between intracellular and extracellular fluid. Nevertheless, urea has to be excreted and this is accomplished most effectively at high urine flows (i.e. when plasma ADH level is low) and there is little urea reabsorption in the collecting tubules.

Urea is the end-product of protein metabolism; the pyrimidines of nucleic acids are also metabolized to urea. Purines (the other nitrogenous constituents of nucleic acids) are metabolized to xanthine and, in most mammals, on to allantoin (which, like urea, is non-toxic and highly water-soluble). However, man and other primates lack the liver enzyme uricase, which converts uric acid into allantoin. Hence the kidneys excrete uric acid, via the organic acid secretory pathway (p. 66).

Some ammonia is excreted in the urine, but most of this is generated within the kidneys (pp. 131–133).

Disease conditions **14**
which alter renal sodium
and water reabsorption

14.1 *Introduction*

A number of disease states alter the renal handling of ions and water. This chapter covers the renal handling of ions (mainly sodium) and water in the following conditions:

(1) congestive heart failure;
(2) shock;
(3) hypertension;
(4) liver disease;
(5) nephrotic syndrome.

In all except the last of these conditions (and possibly also in some forms of hypertension), altered renal function can be regarded as essentially a *compensatory* response, to maintain effective circulating volume. Before considering these disorders, it is necessary to consider the factors which can alter the balance between the formation and reabsorption of tissue fluid.

14.2 *Oedema*

Oedema is an increase in the interstitial fluid volume, so that swelling of the tissue results. There are many clinical states which lead to oedema, but the immediate cause is always a change in the rate of formation or reabsorption of tissue fluid, such that the rate of formation exceeds the rate of reabsorption. The formation of tissue fluid depends on Starling's forces (Chapter 1), so that

Net formation of tissue fluid
$\propto$ [Forces favouring filtration out of capillary]
$-$[Forces opposing filtration out of capillary]

$$\propto (P_{cap} + \Pi_{if}) - (\Pi_{cap} + P_{if})$$

(where P_{cap} = capillary hydrostatic pressure; P_{if} = interstitial fluid hydrostatic pressure; Π_{cap} = capillary oncotic pressure; Π_{if} = interstitial fluid oncotic pressure). A change in any of these four forces can lead to oedema.

The capillary hydrostatic pressure (P_{cap}) is markedly influenced by alterations in venous pressure. The arterioles, which have a high resistance to flow, ensure that only a relatively small pressure is transmitted from the arteries to the capillaries and that changes in arterial pressure have little effect on capillary hydrostatic pressure. However, there are no high-resistance vessels between the capillaries and the venous side of the circulation and consequently changes in venous pressure can have a considerable effect on the capillary hydrostatic pressure.

Increases in capillary hydrostatic pressure will occur, therefore, in any circumstances which increase the venous pressure, e.g. during volume expansion (when an increased intravascular volume increases venous pressure) or when venous return is inadequate (e.g. prolonged standing causes venous pooling in the legs and leads to oedema of the feet and ankles) or when the venous return is obstructed.

The plasma protein osmotic pressure (oncotic pressure, Π_{cap}) opposes the capillary hydrostatic pressure and is responsible for the reabsorption of tissue fluid into the vascular system at the venous end of the capillaries. Volume expansion (e.g. by renal sodium retention), in addition to increasing the capillary hydrostatic pressure (see above), will also tend to dilute the plasma proteins, thereby reducing the force for tissue fluid reabsorption and exacerbating the oedema. Π_{cap} can also be reduced by albumin loss in the urine or as a result of decreased plasma protein synthesis by the liver.

Interstitial fluid oncotic pressure (Π_{if}) depends on the protein content of the interstitial fluid, and the capillaries have a very low permeability to plasma proteins. However, small quantities (mainly albumin) do seep across and, in some circumstances, proteins accumulate in the interstitial fluid and cause oedema. This occurs in conditions of: (a) increased capillary permeability; and (b) obstructed lymphatic drainage – the plasma proteins which leak out of the capillaries normally enter the lymphatic system and are eventually returned to the vascular system via the thoracic duct. Blockage of the lymphatics, e.g. by a tumour, or by parasites (as in elephantiasis), increases the interstitial oncotic pressure and so causes oedema.

The interstitial fluid hydrostatic pressure (or 'tissue turgor pressure') is very difficult to measure accurately, but is close to 0 mmHg.

Local oedema can occur without an overall increase in body fluid content, but, in generalized oedema, body fluid volume is increased.

14.3 *Congestive heart failure*

When the cardiovascular system is failing to provide normal perfusion of the tissues, renal functional adjustments occur which can be regarded as adaptations to increase the effective circulating volume. Such adjustments occur in congestive heart failure. In this condition, the reduction in cardiac output reduces renal perfusion and the effect to the kidney is as if there is hypovolaemia. Here we see the reason for using the term 'effective circulating volume' rather than just 'circulating volume' or 'vascular volume'. In chronic congestive heart failure the vascular volume is normal (or may be elevated) but the effective circulating volume is reduced.

The diminished effective circulating volume has the same effect at the kidney as true hypovolaemia – i.e. it promotes the renal retention of NaCl and water (for mechanism, see p. 111). The extent of this retention will depend on the severity of the heart failure, but if the failure is marked, the blood volume expands so much that the limit of venous distensibility is reached and venous pressure then increases. At the same time, the retention of NaCl and water dilutes the plasma proteins and hence reduces the plasma protein osmotic pressure. These two factors (increased venous pressure leading to increased capillary pressure, and decreased plasma protein osmotic pressure) give rise to oedema.

However, the oedema is essentially a side effect of a **compensatory renal response**, which increases effective circulating volume. The way in which the increased effective circulating volume is brought about is shown in Figure 14.1 (Starling curves).

It can be seen from the figure that expansion of the vascular volume restores tissue perfusion, by increasing the left-ventricular end-diastolic pressure. However, this may lead to a 'side effect' which outweighs the advantage of the restored tissue perfusion. This side effect is pulmonary oedema. If the left-ventricular end-diastolic pressure rises dramatically, there is transmission of this increased pressure back into the lungs – i.e. left atrial pressure is increased, as is pulmonary venous pressure, and the pulmonary capillary pressure, leading to pulmonary oedema. In the lungs, tissue fluid formation should not occur (otherwise the alveoli fill with fluid and become ineffective), so in health the forces for reabsorption of tissue fluid are greater than those for formation of tissue fluid, throughout the length of the pulmonary capillaries.

From the foregoing, it is apparent that the oedema of congestive heart failure occurs as a consequence of the renal retention of NaCl and water, to increase the effective circulating volume. This response of the kidney becomes counterproductive if pulmonary oedema occurs. Since it is the heart which is malfunctioning, treatment should logically be directed at the heart – i.e. treatment with digitalis or similar drugs – to restore a more normal cardiac output, although there is no doubt that diuretics do relieve the symptoms of congestive heart

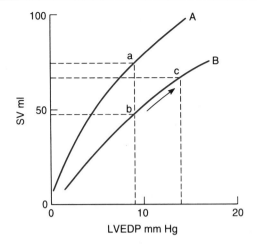

Figure 14.1 The way in which an increase in effective circulating volume is brought about by the kidney in congestive heart failure. An important determinant of cardiac output is the filling pressure (LVEDP, left-ventricular end-diastolic pressure). The normal relationship of LVEDP to stroke volume (SV) is shown by curve A. Point (a) on this line represents tyical normal figures for LVEDP and SV, i.e. at the LVEDP of 9 mmHg, SV is 75 ml. In heart failure, the effectiveness of the cardiac contraction is reduced, and a new relationship between LVEDP and SV is established (curve B). If there is no change in the filling pressure, the stroke volume will be reduced to less than 50 ml (point b). This reduces cardiac output and decreases the effective circulating volume; the renal response to this is fluid and water retention, which increases the central venous pressure, which in turn increases the right ventricular filling pressure so that more blood is expelled by the right side of the heart, in turn increasing the LVEDP and so increasing the stroke volume to the value at point (c). (Similar curves can be drawn relating right-side SV and RVEDP (right-ventricular end-diastolic pressure), but normal RVEDP is approximately 5 mmHg less than LVEDP.)

failure. However, it should be remembered that systemic oedema is of no danger to the patient and occurs as part of a compensatory mechanism. Only pulmonary oedema calls for urgent treatment with diuretics to reduce the body fluid volume. Diuretics are definitely advantageous when they reduce pulmonary congestion, but they also reduce the effective circulating volume.

The tissue anoxia resulting from circulatory inadequacy in congestive heart failure may cause potassium to leak out of cells and lead to a low total body potassium content before any drug therapy is started. Diuretics may then exacerbate potassium depletion. Furthermore, potassium depletion may occur without being apparent because the plasma potassium concentration can be normal (i.e. the loss is from the cells) and this is important because cellular potassium depletion increases the risk of digitalis-induced arrhythmias.

14.4 *Hypovolaemia and shock*

Decreases in extracellular fluid volume produced by fluid loss lead to a reduction in cardiac output (due to reduced end-diastolic pressure, Figure 14.1) and, consequently, tissue perfusion is also reduced. The distinction between hypovolaemia and shock is essentially one of degree. Thus the donation of 500 ml of blood produces hypovolaemia, but the loss of 1 litre of blood (20% of the blood volume), in addition to hypovolaemia, produces mild shock.

Shock is a life-threatening state with a marked reduction of cardiac output and inadequate perfusion of most organs. Hypovolaemia and mild shock cause tiredness, thirst and dizziness. More severe falls in effective circulating volume are accompanied by signs of increased sympathetic activity (tachycardia, pallor, sweating) and impaired function of vital organs (confusion or coma due to cerebral ischaemia, oliguria (low urine flow) or anuria (no urine flow) and acid–base disturbances due to impaired renal function). There are types of shock in which there is not a reduced blood volume, but all types of shock lead to a reduction in effective circulating volume.

The different mechanisms involved can be grouped into two types, hypovolaemic and cardiogenic.

Hypovolaemic shock

In this type of shock, central venous pressure (CVP) decreases, venous return is reduced and hence cardiac output falls. There are three common causes of hypovolaemic shock:

(a) loss of blood – haemorrhage;
(b) loss of plasma – burns;
(c) loss of fluid – persistent vomiting, severe gastroenteritis, excessive sweating.

Cardiogenic shock

Sudden reductions in cardiac output, e.g. due to myocardial infarction, do not change the intravascular volume, but the venous pressure is increased. This change in venous pressure is in the same direction as occurs in the chronic condition of congestive heart failure. However, the increased central venous pressure (CVP) of cardiogenic shock is a direct consequence of the inability of the heart to pump blood adequately (so right-ventricular end-diastolic pressure is elevated), whereas in congestive heart failure the increased CVP is more likely to be a consequence of the renal response to inadequate perfusion.

In both types of shock, both the effective circulating volume and the blood pressure are reduced. In hypovolaemic shock, the decreased effective circulating volume is due to a decreased ECF volume. In cardiogenic shock, the decreased effective circulating volume is due to inadequate circulation, although the intravascular volume is normal. Exactly what effects the reduction in blood pressure will have on renal function will depend on the magnitude of

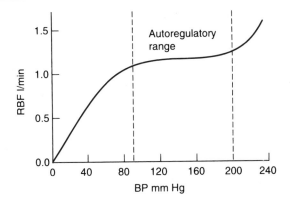

Figure 14.2 Autoregulation of renal blood flow. Renal blood flow (RBF) is plotted against mean arterial blood pressure (BP). In the pressure range 90–200 mmHg, changes in BP have little effect on RBF (but see Chapter 7). Below a BP of 80 mmHg, RBF falls markedly.

the reduction. Figure 14.2 shows the autoregulation of renal blood flow (see also p. 91). However, autoregulation means that the kidney is capable of automatic adjustments of its vascular resistance: it does *not* mean that extrinsic influences do not change renal blood flow. When the effective circulating volume is reduced, blood pressure is maintained by increased sympathetic nervous activity (via baroreceptor reflexes), which causes vasoconstriction in most parts of the body (except the brain), including the kidney. Thus the sympathetic efferent nerves to the renal arterioles (primarily to afferent arterioles) can override the autoregulatory mechanism and lower the renal blood flow. Vasoconstrictor stimuli to the kidney, however, lead to increased renal cortical synthesis of vasodilator prostaglandins (PGE_2 and PGI_2), so that, generally, the renal blood flow remains adequate (and in addition, efferent arteriolar vasoconstriction occurs to maintain filtration pressure) for glomerular filtration, unless the mean blood pressure falls below about 80 mmHg.

Below a mean blood pressure of 80 mmHg, the renal blood flow falls drastically, glomerular filtration decreases, and renal function is impaired. Unless there is prompt restoration of the effective circulating volume, there is the danger of acute renal failure (see box p. 174). Let us consider in detail the renal effects of shock, using hypovolaemic shock caused by haemorrhage as an example.

14.4.1 *Haemorrhage*

Rapid loss of 20% of circulating blood volume produces compensated mild shock. If 30% of the blood volume is lost, there is moderately severe shock (systolic blood pressure below 90 mmHg, heart rate over 90 beats/min),

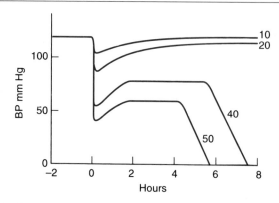

Figure 14.3   The effects of losses of 10, 20, 40 and 50% of the circulating blood volume on systolic blood pressure (BP). Losses of up to 20% of the circulating blood volume have little effect on the blood pressure, although increased sympathetic nervous activity is necessary in order to maintain BP. A loss of **40%** of the circulating blood volume lowers blood pressure dramatically, and although increased sympathetic activity can reduce the fall in BP, the degree of vasoconstriction necessary has serious consequences which may lead to death (see text). Larger haemorrhages (50% of circulating blood volume) are more rapidly fatal.

40% loss causes severe shock, fatal if untreated, and a loss of over 50% is rapidly fatal (Figure 14.3).

Initially, we will consider the 'compensatory phase' (i.e. where the loss is of less than 20% of the circulating blood volume and the systolic blood pressure remains above 90 mmHg). In such cases, there is increased sympathetic nervous activity, leading to tachycardia, increased myocardial contractility and peripheral vasoconstriction. These changes serve to maintain blood pressure, since

$$\text{Blood pressure} = \text{Cardiac output} \times \text{peripheral resistance}$$

and

$$\text{Cardiac output} = \text{Heart rate} \times \text{stroke volume}$$

(Stroke volume is determined by end-diastolic volume and contractility, so that increasing myocardial contractility will increase stroke volume even if end-diastolic volume is constant.)

These actions of the sympathetic nervous system, serving to maintain blood pressure, are beneficial in the short term, but can have deleterious effects if prolonged. This is because excessive tachycardia shortens diastolic filling time and impairs coronary blood flow (coronary blood flow is greater during diastole), and prolonged vasoconstriction has the following adverse consequences:

(1)   Increased aggregation of blood cells, and consequent increased viscosity and microembolization.

(2) Hypoxia in the gastrointestinal tract leading to (a) increased fragility of lysosomes (and an increased concentration of lysosomal enzymes in the plasma), and (b) the production of a factor depressing the activity of the reticuloendothelial system, so that, within two hours of haemorrhage, the body's ability to destroy bacteria or to break down endotoxins is considerably reduced.

(3) Hypoxia in the liver, leading to failure of glycogenesis and failure to convert protein metabolites into urea. Hypoglycaemia, increased blood lactate levels and acidosis occur.

(4) Hypoxia in muscle, leading to lactic acidaemia and increased plasma potassium concentration, which has an adverse effect on cardiac performance.

(5) Hypoxia in the heart, decreasing myocardial performance and exacerbating hypoxia in the rest of the body.

(6) Modified renal function (see below).

14.4.2 *Effect of haemorrhage on body fluid volume and composition, and renal function*

The problems associated with haemorrhage can be divided into two groups:

(1) those due to volume depletion leading to inadequate tissue perfusion;

(2) those due to the electrolyte and acid–base disorders which can accompany volume depletion.

In haemorrhage, the fluid (blood) which is lost is isotonic and thus there is no direct effect on the osmoreceptors. However, haemorrhage increases ADH release via the volume receptors so that water is retained (i.e. osmoregulation becomes subordinated to volume regulation, Chapter 9).

The main change in renal function after haemorrhage is increased sodium reabsorption. Indeed, increased sodium reabsorption (and hence decreased sodium excretion) is such a characteristic feature of hypovolaemia that it can be used diagnostically. The mechanisms responsible for this decreased sodium excretion have been considered in Chapter 9, but will be briefly reiterated here. Haemorrhage reduces effective circulating volume which then reduces renal perfusion pressure. However, renal sodium conservation occurs before decreases in GFR can be observed, indicating changes in tubular reabsorption. GFR can be maintained in spite of some degree of afferent arteriolar constriction, if the efferent arterioles are also constricted (i.e. there is increased filtration fraction). The peritubular capillary hydrostatic pressure, however, will be reduced, favouring increased sodium reabsorption. In addition, the decreased renal perfusion pressure decreases afferent arteriolar wall tension and so stimulates the release of renin and the production of angiotensin II which increases adrenal aldosterone release to promote sodium retention. The intrarenal actions of angiotensin II also favour sodium reabsorption (Chapter 9).

Renal failure

If renal glomerular filtration decreases from the normal 120 ml/min to levels of 30 ml/min or less, symptoms of renal failure become apparent, as a consequence of the retention of nitrogenous waste products (mainly urea and ammonia), and water and electrolyte retention. In fact, renal failure is generally defined as a fall in glomerular filtration leading to increased plasma urea levels. The condition may be *chronic,* with progressive loss of functioning nephrons leading to a gradual decline in renal function, or *acute.* Acute renal failure is usually due to renal ischaemia and hypoxia, or to toxic drugs, or urinary obstruction.

Acute renal failure

The renal regulation of the internal environment depends on (a) ultrafiltration of the plasma, (b) the normal functioning of the tubules to selectively reabsorb and secrete, and (c) the excretion of urine via the ureters, bladder and urethra. Disorders in these mechanisms can lead respectively to **pre-renal failure** (if the renal perfusion is inadequate), **intrinsic renal failure** if the kidneys are damaged, or **post-renal failure** if there is an obstruction in the urinary tract.

Pre-renal failure

Pre-renal failure is failure of the systemic circulation to supply the kidneys with an adequate blood flow to maintain GFR. It may be caused by congestive heart failure, or any of the other conditions associated with decreased effective circulating volume (see text). Essentially, renal function is appropriate in pre-renal failure (i.e. avid Na^+ and water reabsorption occurs, as described in the text). Restoration of the effective circulating volume restores normal renal function and GFR.

Intrinsic renal failure

Intrinsic renal failure usually involves acute tubular necrosis. It can develop from pre-renal failure if the ischaemia persists. Restoration of effective circulating volume does not restore normal GFR in intrinsic renal failure and the excretory function of the kidney is severely impaired.

The increased ADH release brought about by the 'volume' receptors in the atria and by the baroreceptors in the arteries, leads to water retention, so that plasma osmolality decreases (Chapter 9); since the main solute in the plasma is Na^+, this decreased osmolality represents a decreased $[Na^+]$, which therefore acts as a direct stimulus in the adrenal cortex to increase aldosterone release. Extrarenal receptors also modify renal function: decreases in systemic blood pressure reflexly (via the baroreceptors) increase renal sympathetic nerve activity and this is an additional stimulus to renin release.

In concert, the above mechanisms can reduce the urinary sodium concentration to less than 1 mmol/l. Such effective sodium reabsorption can lead to disturbances of acid-base balance because the reabsorption and excretion of other ions are influenced by sodium reabsorption. The sodium concentration in the glomerular filtrate is 140 mM, whereas the filtrate Cl^- concentration is only about 110 mM. Thus from each litre of glomerular filtrate, only 110 mmol Na^+ can be reabsorbed with Cl^- following to maintain electroneutrality. Any additional sodium reabsorption must involve other ways of maintaining electroneutrality, i.e. H^+ and K^+ secretion. Thus

$$Na^+ \text{ reabsorption} \equiv Cl^- \text{ reabsorption} + H^+ \text{ secretion} + K^+ \text{ secretion}$$

So, when sodium is being maximally conserved, K^+ and H^+ are lost from the body and the kidneys are tending to produce *metabolic alkalosis*. However, since in hypovolaemia or other forms of shock, other tissues are inadequately perfused and therefore are hypoxic, anaerobic metabolism leading to acid production may be occurring (i.e. metabolic acidosis), so that there may be little overall change in acid–base balance. This can change dramatically to severe metabolic acidosis when the kidneys can no longer maintain H^+ secretion (intrinsic renal failure, see box and below).

Additional problems of severe haemorrhage (hypovolaemic shock)
As the degree of hypovolaemia increases, there occurs a stage at which tissue perfusion becomes inadequate. When the perfusion of the kidneys is not sufficient for the maintenance of normal urine flow, H^+ secretion can no longer occur at an adequate rate and **metabolic acidosis** can occur. This is exacerbated by inadequate blood flow to other tissues (e.g. muscle), so that tissue respiration becomes partially anaerobic and lactic acidosis occurs. Restoration of perfusion will correct such acidosis, but HCO_3^- should be administered if arterial pH is below 7.2.

Measures to prevent irreversible renal damage
In severe volume depletion the stimulus for renal vasoconstriction is so intense that the renal blood flow may not be restored by measures (e.g. blood transfusion) to restore the circulatory volume, and renal failure can occur due to anoxia and necrosis.

14.5 *Hypertension*

The kidneys contribute to the regulation of blood pressure by regulating the extracellular fluid volume and by releasing vasoactive substances (hormones) into the blood.

In Chapter 9, the renal regulation of ECF volume was discussed. The blood volume is determined by the ECF volume and, since blood volume influences cardiac output, which in turn influences blood pressure, it is clear that ECF volume is a potential determinant of blood pressure. However, changes in blood volume and the consequent changes in blood pressure are normally accompanied by compensatory changes in renal sodium and water excretion, so the persistence of a high blood pressure (i.e. hypertension) may indicate the presence of a disturbance in the kidneys' response to the increased pressure.

In some types of hypertension (secondary hypertension) the causes of this abnormal responsiveness of the kidney are known. These include renal artery stenosis (renovascular hypertension), intrinsic renal disease (renal hypertension), primary hyperaldosteronism (leading to renal sodium retention) or excessive renin production (e.g. in some renal tumours).

14.5.1 *Secondary hypertension*

Renovascular hypertension
This is caused by the renal response to reduced renal perfusion (e.g. due to the stenosis of the renal artery or of one of the interlobar arteries). Reduced perfusion of the afferent arterioles stimulates renin release from the juxtaglomerular apparatus, increasing the production of angiotensin II and thereby causing increased blood pressure both directly (via the vasoconstrictor action of angiotensin II) and indirectly (via salt and water retention brought about by angiotensin and aldosterone).

Renal hypertension
Impaired renal excretion leads to extracellular volume expansion, which can lead to hypertension. In addition, the kidney cortex and medulla synthesize vasodepressor prostaglandins and, although these are thought to have predominantly intrarenal functions, they may also have a systemic role in maintaining normotension, so that inadequate renal prostaglandin synthesis could lead to hypertension.

Primary hyperaldosteronism
This is a rare condition, accounting for less than 1% of hypertension cases. Excessive aldosterone release by the adrenal cortex (usually as a result of an adrenal cortical adenoma) promotes distal nephron sodium reabsorption and potassium secretion. There is normally little or no volume expansion because

of 'escape' from the sodium retaining actions of aldosterone (p. 109). However, there is continued potassium loss and most patients (90%) with primary hyperaldosteronism have hypokalaemia (with plasma $[K^+]$ less than 3.5 mmol/l). Metabolic alkalosis ($[HCO_3^-]$ greater than 30 mmol/l) may also be present, since H^+ as well as K^+ is lost from the distal nephron. It is not entirely clear why the condition produces hypertension; the degree of volume expansion is seldom large enough to account for the elevated blood pressure. The diagnosis of primary hyperaldosteronism is based on the observations of hypokalaemia, a high rate of aldosterone excretion and low plasma renin activity.

14.5.2 *Essential (primary) hypertension*

In most hypertensive patients there is no obvious cause of hypertension and this condition is termed essential hypertension. Renal function is usually disturbed (i.e. the kidney is not responding to the hypertension with increased salt and water excretion), but it remains unclear whether the condition is caused by abnormal renal function or whether it causes abnormal renal function. Figure 14.4 is a scheme showing the interrelationship of renal function and blood pressure.

About 90% of all hypertensive subjects have 'benign essential hypertension', so called because the condition worsens only gradually. Pathological changes in the kidney in this condition include hypertrophy of the tunica media of the renal afferent arterioles with consequent narrowing of the vascular lumen (arteriolar nephrosclerosis). Larger vessels (arteries) may also be affected (arteriosclerosis).

In some patients, the hypertension becomes rapidly progressive (malignant hypertension), characterized by extremely high blood pressures (exceeding 230/130 mmHg), with spontaneous haemorrhages and impairment of the renal blood flow. There is fibrinoid necrosis of the arteriolar walls of many organs, including the kidneys.

Malignant hypertension is almost invariably associated with very high levels of plasma renin, but it is not clear at present whether this is a cause or simply a result of the hypertension and impaired renal blood flow. In benign essential hypertension (even in subjects who subsequently develop malignant hypertension), there are no consistent changes in plasma renin activity.

Recently, it has been suggested that changes in cell-membrane ion-transport processes could be associated with essential hypertension. Specifically, it has been proposed that if the cellular extrusion of Na^+ were reduced, then because intracellular Na^+ would increase, the gradient for Na^+ entry would be reduced, and this would reduce Na^+/Ca^{2+} exchange (counter-transport), resulting in an increased intracellular Ca^{2+}. Such an increase occurring in

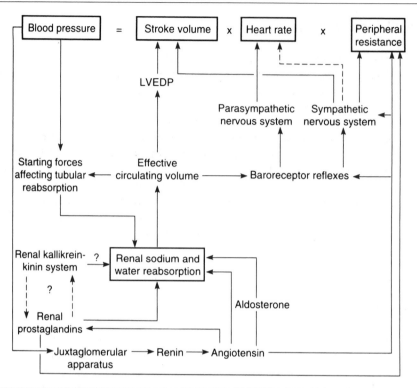

Figure 14.4 Renal involvement in the regulation of blood pressure. LVEDP – left-ventricular end-diastolic pressure. Alterations of blood pressure lead to changes in the release of renin, angiotensin, aldosterone, prostaglandins and kinins. Starling forces affecting tubular absorption are also changed. The modifications to renal function, to regulate the effective circulating volume, may then alter stroke volume, heart rate and peripheral resistance, which determine the blood pressure. Directional changes (i.e. increases or decreases) are not shown on the diagram, but will be clear from the text and from Chapter 9. Atrial natriuretic peptide is not shown, as its functional importance is still not clear.

blood vessels would increase vascular contractility. A related hypothesis is that the alteration of intracellular ion composition stems from a change in the composition of cell-membrane lipids, which alters ionic permeability.

In summary, if systemic arterial pressure is high, but natriuresis and diuresis are not occurring, then the normal relationship between afferent arteriolar pressure and diuresis does not apply. Has the altered pressure-natriuresis relationship caused hypertension, or occurred as a consequence of it? This question cannot be answered at present.

14.6 *Liver disease*

It is a common clinical observation that patients exhibiting symptoms of liver disease – such as jaundice, ascites or portal venous hypertension – frequently develop oliguria (reduced urine flow), sodium retention or other symptoms of disordered renal function.

How does liver disease lead to disorders of kidney function? The accumulation of oedema fluid in the peritoneal cavity is termed **ascites** and frequently occurs when there is more general oedema (e.g. in heart failure), but it is a particular feature of abnormal liver function, as its usual cause is an increased hydrostatic pressure in the hepatic portal vein. The hydrostatic pressure in this vessel increases when there is an obstruction within the liver and the raised pressure forces fluid out of the intestinal capillaries into the peritoneal cavity. It is also possible for ascites fluid to form by transudation from the sinusoids within the liver if the hepatic vein is obstructed.

Renal function in pregnancy

GFR increases markedly (typically by about 50%) in the first trimester of pregnancy and remains elevated until the final month of gestation.

Renal blood flow also increases during the first two trimesters and can be 80% above the non-pregnant level. Near term, renal blood flow begins to decline, but remains well above the non-pregnant level.

The mechanisms responsible for the changes in renal haemodynamics in pregnancy are not clear. However, the changes have important consequences. Production of creatinine and other nitrogenous waste products is not greatly increased in pregnancy, so the increased GFR leads to reduced plasma concentrations of these substances. The increased filtered loads of glucose and amino acids can also lead to these being present in urine, as there is not a corresponding increase in tubular reabsorptive capacity to match the increased GFR. In fact, there are reports that tubular glucose reabsorptive capacity decreases.

Plasma osmolality decreases during the early months of pregnancy, and reaches its lowest value (275 mosmol/kg H_2O compared to the normal level of 285) in the third month. The decrease is primarily due to reduced plasma sodium concentration; the decrease does not suppress ADH release, i.e. the osmotic threshold for the regulation of ADH is altered.

The secretion of aldosterone into the plasma from the adrenal cortex is increased in pregnancy, but the effect of this is largely counterbalanced by progesterone, synthesized in increased amounts in pregnancy, which antagonizes the sodium-retaining effect of aldosterone.

A potentiating factor in the development of ascites in liver disease is decreased albumin synthesis (the liver is the source of plasma albumin), so that the plasma protein osmotic pressure decreases. The loss of part of the circulating volume by transudation from the capillaries into the peritoneal cavity decreases the effective circulating volume and leads to a renal compensatory response – increased NaCl and water reabsorption. A proportion of the fluid thus retained itself becomes ascites.

As the ascites develops, the intra-abdominal pressure rises and raises the venous pressure in the veins which pass through the abdomen. Thus, the venous drainage of the lower limbs becomes impaired and oedema of the lower extremities develops. Patients with ascites may also have arteriovenous fistulas within the liver, so that the effective circulating volume is further reduced (since blood passing directly from arteries to veins is not effectively perfusing the tissues).

The development of ascites in liver disease is shown in Figure 14.5. The cautionary remarks concerning the use of diuretics in congestive heart failure also apply to their use in hepatic disease. There is no logical reason why

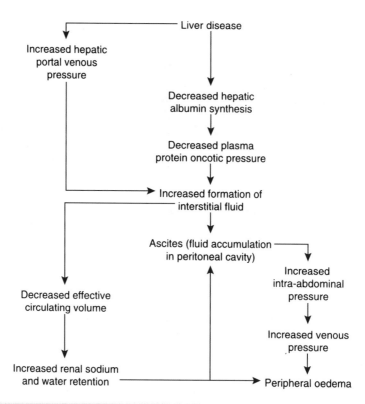

Figure 14.5   The development of ascites in liver disease.

ascites fluid needs to be rapidly removed. Its removal should be a gradual process, since part of the accumulation was in response to renal compensatory mechanisms maintaining effective circulating volume. The rapid removal of ascites fluid by a powerful diuretic could catastrophically decrease the effective circulating volume and in addition produce potassium loss. Electrolyte disturbances in the presence of liver disease can be serious and may lead to coma and death (hepatic coma), commonly due to excessively high levels of ammonia in the blood. The sequence of events is as follows: potassium loss from the body leads to hypokalaemia, and the low extracellular fluid $[K^+]$ causes K^+ to leave the cells and Na^+ and H^+ to move in. When this occurs at the renal tubule cells, the lowered cellular pH stimulates NH_3 production from glutamine. Although much of the ammonia so formed enters the renal tubules and is excreted, some enters the blood and, if the liver is healthy, is converted into urea by the liver. In the presence of a diseased liver, the ammonia produced by the kidneys increases the blood ammonia concentration.

Another way in which liver disease (e.g. cirrhosis) complicates renal function is due to the fact that the liver plays a considerable part in the inactivation of circulating ADH and aldosterone, and this inactivation is less effective in cirrhosis.

14.7 *Nephrotic syndrome*

In this syndrome, the glomerular filtration barrier becomes permeable to plasma proteins and consequently there is proteinuria, with a progressive reduction in the plasma protein osmotic pressure (Π_{cap}). Albumin is the smallest plasma protein and therefore is filtered most readily in the nephrotic syndrome, and it is also the protein which contributes most to the plasma protein osmotic pressure. Thus it is the fall in albumin concentration which is the main cause of the decreased plasma protein osmotic pressure. This in turn alters the 'Starling forces' across the capillaries and causes oedema (Figure 14.6).

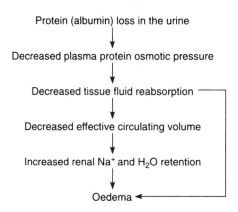

Figure 14.6 Oedema in the nephrotic syndrome.

The reduced effective circulating volume leads to renal sodium and water retention by the same mechanisms as the ones which occur in congestive heart failure.

Further reading

Levy, M. (1992) Hepatorenal syndrome. In D. W. Seldin and C. Giebisch (eds), *The kidney, physiology and pathophysiology,* 2nd edn, Raven Press, New York, pp. 3305–3326

Lote, C. J., Harper, L. and Savage, C. O. S. (1996) Mechanisms of acute renal failure. *Brit. J. Anaesthesia* **77**, 82–89

Use of diuretics $\quad$ *15*

15.1 *Introduction*

Strictly, diuresis is increased urine flow and diuretics are substances which elicit diuresis. By this definition, water is a diuretic. In medicine, however, 'diuretic' has come to have a more specific meaning. In the kidney, water reabsorption is dependent primarily on Na^+ reabsorption and the term diuretic is generally used to mean an agent which inhibits tubular Na^+ reabsorption. With the exception of osmotic diuretics (see below), most diuretics are organic acids and hence are secreted by the proximal tubule organic acid secretory mechanism and exert their effects from within the tubule.

There are a number of different chemical types, and there are several possible sites of action for diuretics within the nephron.

15.2 *Osmotic diuretics*

If the blood glucose concentration rises so that the nephrons are unable to reabsorb all the filtered glucose, then glucose is excreted in the urine. In addition, the urine flow increases, i.e. diabetes mellitus is usually characterized by the excretion of a large volume of urine as well as by glucose excretion. Some sodium chloride is also lost and the urinary osmolality is close to plasma osmolality. This is an osmotic diuresis.

Osmotic diuresis can be induced by the intravenous administration of a non-reabsorbable solute, such as mannitol. The reduction in water (and NaCl) reabsorption which occurs in osmotic diuresis takes place mainly in the proximal tubule and loop of Henle. In the proximal tubule, the normal reabsorptive process does not alter the osmolality of the tubular fluid, which remains at about 285 mosmol/kg H_2O (almost isotonic to plasma). Furthermore, the reabsorptive process does not normally change the tubular sodium concentration, since water absorption follows solute absorption, and hence water is reabsorbed at a rate which maintains a constant sodium concentration.

The presence of non-reabsorbable solute in the proximal tubule limits water reabsorption, because the concentration of such a solute will increase as water is absorbed, and this increased non-reabsorbable solute concentration will osmotically oppose further water reabsorption. Sodium reabsorption then lowers the tubular sodium concentration, until the gradient for the passive backdiffusion of sodium into the tubule is such that net sodium reabsorption ceases.

Another action of osmotic diuretics, the reason for which is still not entirely clear, is an increase in the renal medullary blood flow which leads to a reduction in the corticomedullary solute gradient.

During osmotic diuresis, the renal oxygen consumption is almost unaltered, despite the reduction in sodium reabsorption. This is probably due to the fact that the rate of active Na^+ transport is unchanged (but backflux of Na^+ into the tubular lumen is increased). Osmotic diuretics do not normally increase sodium excretion to more than 10% of the filtered load.

15.3 *Loop diuretics*

Most of the recently introduced diuretics primarily affect the loop of Henle, and so are termed loop diuretics. They include ethacrynic acid, furosemide, bumetanide and piretanide. They act primarily by inhibiting NaCl extrusion from the thick segment of the medullary ascending limb, by blocking the $Na^+/2Cl^-/K^+$ cotransporter in the apical membrane of the thick ascending limb cells, and so blocking the entry of these ions into the cells. Loop diuretics get to their site of action on the apical membrane by not only being filtered, but also by being secreted into the nephrons by the organic acid secretory mechanism in the proximal tubules. Since the hypertonicity of the renal medulla depends on the transport of Na^+ and Cl^- out of the thick ascending limbs of the loops of Henle, loop diuretics diminish the corticomedullary concentration gradient, and so reduce the amount of water which can be reabsorbed in the collecting ducts, thereby impairing renal concentrating ability. Furthermore, since it is transport of NaCl out of the thick ascending limb, unaccompanied by water, which generates osmotically free water in the nephrons (see Chapter 8), loop diuretics also diminish the kidneys' ability to produce dilute urine. In high doses, furosemide or bumetanide can result in the excretion of over 30% of the filtered load of sodium and water.

Because the loop diuretics decrease the extracellular fluid volume, they tend to increase aldosterone release (see Chapter 9), and this, together with the increased rate of NaCl delivery into the distal tubule, increases K^+ secretion and excretion.

15.4 *Thiazides*

Thiazide diuretics, such as chlorothiazide, hydrochlorothiazide and benzothia-diazine, are organic anions which reach the tubular lumen in the same way as the loop diuretics, i.e. by a combination of glomerular filtration and proximal tubular secretion. The use of thiazide diuretics has diminished since the introduction of loop diuretics, but they are still used in the treatment of hypertension. The thiazides act on the early distal tubule, where they block the activity of the Na^+/Cl^- cotransporter which transports these ions into the cells from the lumen across the apical membranes. The delivery of sodium to the more distal tubular Na^+/K^+ exchange site is increased, so that K^+ secretion is enhanced. Less NaCl is normally absorbed in the distal tubule than in the medullary ascending limb and consequently the thiazides are less potent than the loop diuretics. They typically produce a diuresis of up to 10% of the filtered Na^+ and water.

Most thiazides have some carbonic anhydrase inhibitory activity and, in addition, they decrease renal blood flow and GFR.

15.5 *Aldosterone antagonists*

Spironolactone competes with aldosterone for receptor sites on the principal cells of the collecting ducts (the structures of the spironolactone and aldosterone are shown in Figure 15.1). As aldosterone promotes absorption of Na^+ and H^+/K^+ secretion, spironolactone causes a natriuresis and reduces urinary H^+ and K^+ excretion. It has little diuretic activity, but its usefulness lies in its ability to reduce the K^+ excretion produced by other diuretics. Thus, a combination of, for example, furosemide (see below) and spironolactone minimizes the disturbance of K^+ balance which would occur with furosemide alone. Aldosterone antagonists (and Na^+ channel blockers, see below) are thus termed potassium-sparing diuretics.

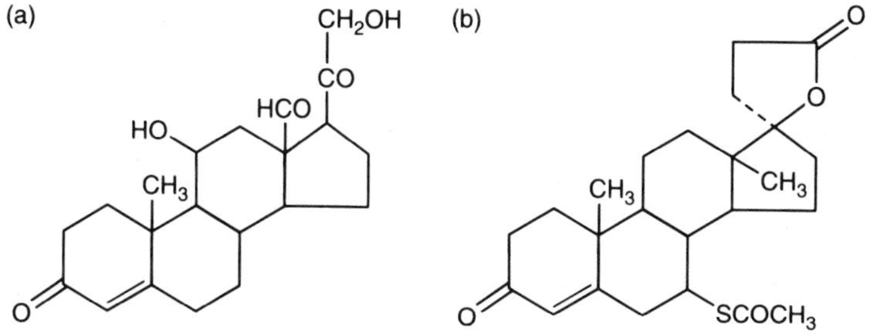

Figure 15.1 Chemical structure of (a) aldosterone and (b) spironolactone.

15.6 *Na channel blockers*

These agents, which include amiloride and triamterene, have a similar effect to spironolactone, i.e. they reduce Na^+ absorption and H^+/K^+ secretion in the distal tubule; but they do this by an action which is independent of aldosterone. They have little diuretic action (causing the excretion of only 2–3% of the filtered load), but like spironolactone are potassium-sparing agents and can be used in conjunction with other diuretics. In low doses, triamterene and amiloride block the entry of Na^+ into the tubule cells across the luminal cell membrane by blocking sodium-selective ion channels, and thus they decrease the availability of Na^+ to the Na^+K^+-ATPase at the basal cell membrane.

15.7 *Carbonic anhydrase inhibitors*

In the proximal tubule, HCO_3^- is absorbed from the tubule by conversion to CO_2, brought about as a result of H^+ secretion, i.e.

$$HCO_3^- + H^+ \rightleftharpoons H_2CO_3 \rightleftharpoons CO_2 + H_2O$$

The second step in this reaction, the conversion of H_2CO_3 to CO_2 and H_2O, is catalysed by carbonic anhydrase, which is present in the brush border area of the proximal tubule cells. Within the cells, carbonic anhydrase is again necessary to reconvert CO_2 to H^+ (which is secreted) and HCO_3^- (which is reabsorbed). So inhibition of carbonic anhydrase inhibits HCO_3^- absorption from the tubule by blocking the reaction sequence (Figure 15.2) in two places.

The presence in the lumen of non-reabsorbable HCO_3^- reduces Na^+ reabsorption, and $NaHCO_3$ passes into the more distal parts of the nephron. In the distal tubule, K^+ secretion is enhanced, mainly because the delivery of Na^+ to this nephron segment is increased, but also due to the fact that, normally, there is some H^+ secretion (in effect, in competition with K^+ secretion) in exchange for Na^+ reabsorption, and the inhibition of carbonic anhydrase reduces the availability of H^+ for secretion.

Carbonic anhydrase inhibitors are weak diuretics, producing a diuresis which seldom exceeds 5% of the filtered load of Na^+ and water. They are used clinically to correct some acid–base disturbances (alkalosis), rather than for their diuretic action. The inhibitor used clinically is acetazolamide ('Diamox').

15.8 *Clinical use of diuretics*

Some of the dangers of diuretic therapy have been mentioned in Chapter 14, the main point being that oedema is almost invariably a symptom of another

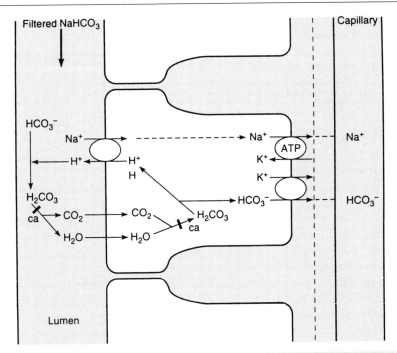

Figure 15.2 Impairment of HCO_3^- reabsorption by carbonic anhydrase inhibitors. The reaction sequence is blocked at the two points shown. ca = carbonic anhydrase.

disorder and not a disease in itself. Furthermore, systemic oedema (as opposed to pulmonary oedema) is not an immediate threat to the life of the patient and there is seldom any justification for the rapid removal of oedema fluid. Indeed there are often good reasons for caution in eliminating oedema rapidly.

In chronic congestive heart failure (Chapter 14), oedema is a consequence of renal fluid retention and this fluid retention is a compensatory response to increase the effective circulating volume. Diuretic therapy to reduce the oedema will also reduce the effective circulating volume, which may, if the heart failure is severe, be barely adequate. Nevertheless, diuretic therapy, if carefully monitored, is generally beneficial in most patients, since some pulmonary congestion is usually present and diuretic therapy reduces this.

Apart from diminution of the effective circulating volume, other disturbances which can be produced by diuretics are as follows.

Acid–base and electrolyte disturbances
Diuretics which cause NaCl excretion (furosemide, ethacrynic acid and thiazides) can produce metabolic alkalosis, because they increase the delivery of NaCl to the distal exchange site where Na^+ is absorbed and K^+ and H^+ are

secreted. Thus both alkalosis and hypokalaemia ensue. The alkalosis is exacerbated by the contraction in body fluid volume brought about by the diuretic, since the HCO_3^- content of the body is essentially unchanged, so that the HCO_3^- concentration increases.

Diuretic agents which act as the distal exchange sites (spironolactone and triamterene) reduce H^+ and K^+ secretion, so their use alone can result in hyperkalaemia and metabolic acidosis. Generally, however, they are used in conjunction with loop diuretics.

Calcium homeostasis and diuretic treatment

In the proximal tubule, calcium reabsorption parallels that of sodium. Consequently, diuretics which reduce proximal sodium and water reabsorption (i.e. osmotic diuretics and carbonic anhydrase inhibitors) decrease proximal calcium absorption and enhance calcium excretion. The fact that carbonic anhydrase inhibitors also alkalinize the urine (and calcium solubility is reduced in alkaline solutions) can lead to the formation of renal stones.

In the ascending thick limb of the loop of Henle, the lumen-positive potential generated by the apical transport of $Na^+/2Cl^-/K^+$ (with some backleak of K^+, so that a net excess of negative ions is removed from the lumen), is a driving force for calcium absorption. Loop diuretics, by reducing the $Na^+/2Cl^-/K^+$ transport, also reduce calcium absorption and so increase calcium excretion. This means that loop diuretics can be used in the treatment of hypercalcemia.

In contrast to the above, thiazide diuretics increase calcium absorption in the distal tubule, and decrease calcium excretion.

Azotaemia and hyperuricaemia

If diuretic administration leads to a reduction in effective circulating volume, then renal perfusion may be reduced, with a consequent fall in urea and creatinine excretion.

Hyperuricaemia can occur because proximal tubular sodium reabsorption determines the extent of uric acid reabsorption. Diuretics with actions distal to the proximal tubule will, by causing volume depletion, actually enhance proximal sodium absorption and uric acid reabsorption will also be enhanced.

Further reading

Agarwal, R. (1997) Use of diuretics in renal disease. In R. L. Jamison and R. Wilkinson (eds), *Nephrology*, Chapman and Hall, London, pp. 883–892

Suki, W. N. and Eknoyan, G. (1992) Physiology of diuretic action. In D. W. Seldin and G. Giebisch (eds), *The Kidney, Physiology and Pathophysiology*, Raven Press, New York, pp. 3629–3670

Answers to problems

Chapter 1

1.1 (a) Ineffective. Of the plasma constituents, only the plasma proteins are effective osmoles across the capillary wall.
(b) Effective. The cell membrane has a low permeability to Na^+ (and the Na^+K^+-ATPase pumps Na^+ out of the cell).

1.2 (a,b) Ineffective. Both capillary walls and cell membranes are highly permeable to urea.

1.3 (a) Intracellular fluid volume would decrease. Water would move osmotically from the cells to the extracellular fluid, hence
(b) The extracellular fluid volume would increase. The shift of water would equalize intracellular and extracellular osmolalities, hence (c,d) both compartments would have a slightly increased osmolality.

1.4 (a,b) After a few minutes the urea, being an 'ineffective' osmole, would have equilibrated between intra- and extracellular fluid, so neither volume would be affected. However (c,d) the additional urea would increase the osmolality of both compartments.

Chapter 3

3.1 (a) Nothing, since the change in K_f will mean that equilibrium still occurs, but a little further along the length of the glomerular capillaries.
(b) The efferent arteriolar oncotic pressure will be reduced, since GFR (and hence the concentrating effect of filtration on the proteins) will be reduced.

3.2 The maintenance of an almost consistent pressure along the glomerular capillaries is due to the presence of the efferent arteriole, a resistance vessel, downstream from the glomerular capillaries. The glomerular capillary pressure is the force which causes glomerular filtration.

189

3.3   The kidneys could not function if the filtration fraction (that fraction of the plasma delivered to the kidneys which is filtered) were very high. Even with a filtration of only 20%, there is an increased viscosity of the blood in the efferent arterioles, as a result of the increased plasma protein concentration and increased haematocrit due to glomerular filtration. If all of the plasma delivered to the kidneys were filtered, the efferent arterioles would be clogged with a sludge of red cells and plasma proteins!

Chapter 5

5.1   (a) Since 110 ml/min of glomerular filtrate is formed, and each ml contains 4.2/1000 mmol glucose, the filtered load is $4.2/1000 \times 110$ mmol/min = 0.46 mmol/min (83 mg/min).

(b) Although trace amounts of glucose are always present in urine, the usual clinical tests for glucose are insufficiently sensitive to detect the glucose excretion of normal subjects. In this subject, with a normal plasma glucose concentration, glucose would not be detectable in the urine by the usual clinical methods.

Chapter 7

7.1   (a) The PAH clearance is given by the formula

$$C_{PAH} = \frac{U_{PAH}V}{P_{PAH}}$$

where U_{PAH} and P_{PAH} are the urine and plasma concentrations of PAH, respectively, and V is the urine flow. The normal units of clearance are ml/min, so, since the concentration terms cancel, we should express the urine flow in ml/min. 180 µl/min is 0.18 ml/min.

U_{PAH} is 1.333 mg/ml, i.e. 1333 µg/ml
P_{PAH} is 20 µg/ml

Thus

$$C_{PAH} = \frac{1333 \times 0.18}{20} = 12 \text{ ml/min}$$

(b) 12 ml/min. PAH clearance is the 'effective' renal plasma flow, as long as the T_m for PAH is not exceeded. The T_m is only reached at a plasma concentration of about 8 mg/100 ml (80 µg/ml).

(c) With a measure of the renal venous PAH concentration, the true renal plasma flow can be calculated from the Fick formula

$$\text{Renal plasma flow} = \frac{U_{PAH}\,V}{P_{APAH} - P_{VPAH}}$$

where P_{APAH} and P_{VPAH} are renal arterial and renal venous plasma PAH concentrations.

$$\text{Renal plasma flow} = \frac{1333 \times 0.18}{20 - 3} = 14.1\,\text{ml/min}$$

(d) Not all of the plasma perfusing the kidney supplies the nephrons. Some goes to the capsule and the fatty tissue. The PAH in this is not available to the tubules for secretion and hence PAH clearance underestimates true renal plasma flow (by 5–10%).

Chapter 10

10.1 The subject has an acidosis (since the arterial pH is below 7.4), with increased ventilation (pCO_2 is 33 mmHg instead of 40 mmHg). Hence the subject has a metabolic acidosis, which has been compensated by increased ventilation.

Index